Rescue Operations Strategies for Saving Lives

Eric J. Neal

Published by Eric J. Neal, 2025.

RESCUE OPERATIONS STRATEGIES FOR SAVING LIVES

First edition. March 17, 2025.

ISBN: 979-8230300960

Written by Eric J. Neal.

Chapter 1: The Foundations of Rescue: Principles, Ethics, and Preparedness

Rescue operations are complex undertakings that require meticulous planning, precise execution, and unwavering commitment to ethical standards. These operations are not only about saving lives but involve a rigorous adherence to principles that ensure both efficiency and safety. At their core, the operations demand an understanding of the dynamic interplay between risk management and the imperative need to act quickly and effectively under often unpredictable circumstances. This complexity is underscored by the necessity for rescuers to navigate not only physical and logistical challenges but also the ethical dimensions inherent in making decisions in high-stakes environments. The ability to balance swift action with thoughtful consideration of all variables is what defines a successful rescue operation.

In this chapter, readers will delve into the foundational elements that underpin effective rescue efforts, focusing on core principles, ethical considerations, and preparedness strategies. It begins by examining the critical commitment to life safety that guides every decision made in the field. Attention is given to scene control, which is vital for managing hazards and ensuring efficient operations. The importance of ethical decision-making is highlighted, including the challenges of maintaining cultural sensitivity and professionalism while remaining empathetic. Furthermore, the chapter explores the strategies employed for risk assessment and safety protocols, emphasizing the role of personal protective equipment and standard operating procedures. Readers will gain insights into the significance of training and preparedness, as well as the integration of modern technologies that enhance rescue capabilities. This exploration provides a comprehensive understanding of how these diverse elements come together to form a cohesive and effective approach to rescue operations.

Introduction to Rescue Operations

A fundamental understanding of rescue operations begins with defining what they are and their primary objectives. Rescue operations involve the organized efforts to save individuals in danger or distress, typically resulting from natural or human-made disasters. These operations aim to ensure life safety, provide medical care, and restore normalcy through the effective allocation of resources and coordination among various emergency services.

One key aspect of successful rescue operations is their structured approach, which significantly impacts saving lives during emergencies. Organized responses minimize chaos and maximize efficiency by establishing clear roles and responsibilities within the participating agencies. In most jurisdictions, emergency response services have priority access to traffic systems, ensuring rapid deployment (Emergency Response Services, n.d.). This operational structure is quintessential as it enables quick decision-making and proper communication, essential factors when time is of the essence.

Historical examples of notable rescues offer valuable lessons that can inform future strategies. For instance, the 2010 Chilean mine rescue demonstrated the importance of international cooperation and technology in a successful outcome. After 33 miners were trapped underground for 69 days, multinational teams worked together using advanced drilling techniques and real-time communications to extract each miner safely. Another significant event was the rescue operation during Hurricane Katrina in 2005, highlighting the need for better coordination and preparedness. The response exposed weaknesses in communication and resource distribution, leading to reforms in disaster management protocols across the United States.

The evolution of modern rescue techniques has played a crucial role in enhancing the efficiency and effectiveness of these operations.

Over the decades, developments in technology have transformed how rescuers approach their tasks. From the use of drones and GPS to sophisticated communication networks, the incorporation of technology allows for precise localization and faster responses. Lakshmi Narayanan & Ibe (2015) stress the necessity of high-precision location information in search and rescue network operations, which narrows the search area and accelerates the rescue process.

Furthermore, specialized resources like Heavy Urban Search and Rescue Teams (HUSAR) and Disaster Medical Assistance Teams (DMAT) have emerged worldwide, equipped with advanced tools to handle complex rescue scenarios. HUSAR teams utilize cutting-edge equipment to explore damaged structures, while DMATs deploy field hospitals in disaster zones, offering critical medical services. Such advancements underscore the increasing specialization and sophistication required in modern rescue missions.

Core Principles and Ethics in Rescue Operations

Life safety stands as the cornerstone of any rescue operation, a principle that demands unwavering dedication both for the individuals in jeopardy and the rescuers risking their own lives to save others. This emphasis on life safety requires a dual focus: ensuring the well-being of victims while simultaneously guaranteeing responder safety, creating an environment where the best possible outcome can be achieved without compromising those involved. To accomplish this, responders must be equipped with proper training and equipment, allowing them to navigate hazardous situations without endangering themselves or exacerbating the situation at hand. Adopting a mindset that prioritizes life safety encourages thorough preparation and continuous vigilance, crucial attributes that underpin successful rescue operations (2025).

Scene control forms the structural backbone of effective rescue efforts, encompassing a range of measures designed to manage hazards, secure the area, and establish command. Scene control begins with hazard management, identifying potential dangers that could impact the rescue operation. This includes assessing physical risks, such as unstable structures or hazardous materials, and implementing strategies to mitigate these threats. By securing the area, unauthorized access is prevented, maintaining order and reducing impediments that may interfere with the operation. Area security also ensures that only trained personnel are present, minimizing the risk of accidents or misunderstandings arising from untrained individuals entering the scene.

Central to scene control is the establishment of a clear command structure, often through the Incident Command System (ICS), which provides a standardized coordination and control approach. ICS enables efficient communication and resource allocation, ensuring that all responding units operate under a unified strategy. The system fosters a collaborative environment where each team member understands their role and responsibilities, enhancing overall effectiveness and reducing the likelihood of chaos in dynamic scenarios. Using ICS in managing rescues facilitates the integration of multiple agencies and disciplines into a cohesive response effort, further underscoring its value in emergency operations (<i>Emergency Responder Experiences with the Incident Command System: A Qualitative Analysis - ProQuest</i>, n.d.).

Ethical considerations in rescue operations extend beyond procedural expertise, demanding an adherence to principles that honor the dignity and rights of all parties involved. Informed consent is a fundamental ethical duty, ensuring that victims understand the nature of the assistance offered and have the autonomy to make decisions about their care. This respect for personal agency is balanced by the need for quick action in life-threatening situations, where obtaining explicit consent might not always be feasible. In such cases, rescuers must exercise sound judgment, guided by ethical principles and professional standards, to act in the best interest of the victim.

Cultural sensitivity is another key component of ethical rescue operations, requiring responders to recognize and respect the diverse backgrounds and beliefs of those they assist. By approaching each individual with empathy and understanding, rescuers can effectively communicate and build trust, facilitating cooperation and ultimately improving outcomes. Cultural competence training can prepare responders to navigate these challenges sensitively and respectfully, reducing potential conflicts and ensuring a more inclusive approach to rescue efforts.

Balancing professionalism with empathy is perhaps one of the most challenging aspects of rescue work, demanding a delicate equilibrium between maintaining objective detachment and expressing compassion towards victims and their families. While professionalism necessitates adherence to protocols and procedures, it should not come at the expense of human connection and support. Showing empathy can provide much-needed comfort and reassurance to those experiencing traumatic events, reinforcing the notion that rescue operations, at their core, are acts of humanity dedicated to preserving life and dignity.

Situational awareness is an integral part of managing rescue operations, involving continuous risk assessment and adaptation to evolving circumstances. This awareness encompasses being alert to changes in the environment, recognizing new hazards, and adjusting tactics accordingly to maintain operational integrity and ensure everyone's safety. Effective situational awareness requires strong observational skills and the ability to process information quickly, enabling rescuers to anticipate challenges and implement solutions proactively.

Understanding Risk and Safety Protocols

Risk assessment and safety protocols are fundamental to ensuring the effectiveness and safety of rescue operations. The dynamic nature of emergencies necessitates a thorough understanding of potential risks and well-established safety measures. This subpoint delves into the importance of these elements, elucidating the types, use, and maintenance of Personal Protective Equipment (PPE); methods for recognizing various hazards; the role of Standard Operating Procedures (SOPs); and the need for safety briefings.

Personal Protective Equipment plays a crucial role in safeguarding rescuers during operations. PPE varies depending on the nature of the operation but typically includes gear such as helmets, gloves, goggles, respiratory protection, and protective clothing. The proper use of PPE involves adhering to guidelines that specify the correct type of equipment for specific situations. For example, firefighting tasks may require flame-resistant clothing and self-contained breathing apparatus, while chemical spills demand specialized protective suits and respirators. Regular maintenance and inspection of PPE are essential to ensure functionality and reliability, as damaged or faulty equipment can compromise safety and effectiveness. Failure to maintain equipment rigorously can lead to significant risks for responders on the field, underscoring its necessity in preparedness.

Recognizing hazards is another critical element in risk assessment. Physical, chemical, and environmental hazards present distinct challenges during rescues. Physical hazards might include unstable structures, debris, and sharp objects. Chemical hazards could stem from substances like toxic gases or corrosive materials, which require swift identification to mitigate exposure risks. Environmental factors—such as inclement weather, wildlife, or geographical obstacles—also pose risks that rescuers must address during operations. Employing dynamic risk assessments allows responders to remain vigilant and adjust their strategies as conditions evolve, enabling them to anticipate and counter potential threats effectively (<i>What Is a Dynamic Risk Assessment? A Guide with Examples</i>, 2024).

Incorporating Standard Operating Procedures ensures consistency and safety in rescue operations. SOPs provide a structured approach to tackling incidents, offering clear instructions on executing tasks while minimizing risks. These procedures encompass actions for handling equipment safely, managing hazardous materials, and coordinating team efforts. Adherence to SOPs helps maintain order and efficiency under pressure, preventing chaos and miscommunication. When SOPs are diligently followed, they enhance the overall performance of the team and reduce the likelihood of accidents or oversights. As a flexible framework, SOPs can be adapted based on the situation's requirements (Staff, 2002).

Safety briefings and pre-rescue assessments form an essential part of preparedness, ensuring teams are fully informed and ready to act. Safety briefings provide an opportunity for leaders to outline objectives, communicate concerns, and explain roles and responsibilities. They emphasize the importance of awareness, encouraging responders to stay alert and responsive. Pre-rescue assessments involve evaluating available resources, identifying potential hazards, and determining necessary actions before engaging in rescue activities. This proactive stance enables rescuers to prepare adequately, optimizing their response to meet any challenges presented during the operation. Conducting these assessments habitually embeds a culture of safety within the team, fostering communication and collaboration.

Furthermore, training in hazard identification and risk assessment techniques empowers rescuers to identify issues quickly and take preventive measures. This knowledge equips them with the skills needed to evaluate risks and implement appropriate control measures. Through continuous education and drills that simulate real-life scenarios, responders enhance their ability to respond effectively. These exercises reinforce theoretical knowledge, allowing practitioners to gain practical experience and familiarize themselves with potential complexities encountered in the field.

Dynamic risk assessment remains a vital tool in developing situational awareness and adaptability among rescue teams. By integrating this approach into routine practices, teams become adept at identifying changes within their environment and adjusting their tactics accordingly. This proactive mindset encourages open communication between team members, facilitating the sharing of critical observations and insights that contribute to a safer workplace. Dynamic risk assessments empower personnel to make informed decisions swiftly, reducing the impact of unforeseen challenges and promoting a culture of continuous improvement (<i>What Is a Dynamic Risk Assessment? A Guide with Examples</i>, 2024).

Preparedness and Training

Preparedness and training are indispensable components that significantly enhance the effectiveness of rescue operations. In high-risk environments where fast decision-making and physical endurance are critical, both physical and mental conditioning of responders is imperative. Responders are often required to work under extreme conditions, such as intense heat, confined spaces, or adverse weather. These conditions necessitate robust physical training akin to that of athletes to ensure strength, stamina, and agility are maintained at peak levels. Furthermore, mental resilience must be cultivated through stress management techniques and psychological readiness programs. The nature of crises can be unpredictable and emotionally taxing, demanding a high degree of mental fortitude from those involved in rescue efforts.

Certification plays a pivotal role in maintaining operational standards in rescue scenarios. It is not only a mark of credibility but also a necessary step in ensuring personnel are proficient in the latest life-saving techniques. Continuous education enables rescuers to stay updated with evolving tools and methodologies which may better address new challenges presented by emergencies. Regular training sessions function as refreshers for existing skills while providing opportunities for learning advanced strategies and technologies. Institutions offering these certifications often collaborate with experts from diverse fields to create comprehensive programs that are both theoretically informative and practically applicable.

Scenario-based drills are one of the most effective methods to prepare teams for real-life situations. These drills simulate potential emergencies, allowing responders to practice coordinated responses in controlled, yet realistic settings. By recreating scenarios such as building collapses, water rescues, or hazardous material spills, individuals and teams can assess their preparedness and refine their tactical approaches. These exercises help identify gaps in coordination or resource allocation, facilitating improvements before actual deployment. For instance, during hurricane response drills, teams practice evacuation procedures and communication protocols, ensuring all members understand their roles in managing chaos and saving lives efficiently.

Technology integration also greatly enhances preparedness efforts in modern rescue operations. Drones, for example, have proven invaluable in providing aerial surveillance, helping locate survivors, and assessing damage in areas that might be inaccessible due to debris or danger. With thermal imaging capabilities, drones can detect heat signatures of people trapped under rubble, enabling faster location of victims even in darkness or smoke-filled environments (Werner, 2015).

GPS technology has revolutionized navigation for rescue teams, ensuring quick and accurate location tracking. This reduces response times as teams can traverse difficult terrain more effectively, directly impacting the efficiency of the rescue attempts. Advanced communication systems, powered by technology, ensure seamless information flow between command centers and field operatives, thereby reducing misunderstandings and enhancing strategic coordination.

Furthermore, these technologies support preemptive measures, allowing for monitoring of disaster-prone areas and facilitating timely evacuations. GIS mapping, integrated with satellite imagery, provides detailed topographical data, helping in planning routes and strategizing deployments in event of a natural disaster (FasterCapital, n.d.). Such technological advancements are vital in developing robust emergency preparedness plans that can adapt swiftly to changing circumstances, ultimately increasing the success rate of rescue missions.

While technology plays a crucial role, it does not replace the need for human expertise and intuition. Rescue operations rely on the empathy and judgment of skilled professionals who can make nuanced decisions in dire situations. Therefore, it's essential that advancements in tools and equipment complement rather than supplant the human element in these operations.

Psychological Resilience and Lessons Learned

In the demanding field of rescue operations, psychological resilience plays a crucial role in ensuring the well-being of first responders and enhancing their capacity to carry out their duties effectively. The work environment for firefighters and other first responders inherently involves high-stress situations that can lead to mental health challenges, such as stress and PTSD. Addressing these issues requires an understanding and implementation of effective coping mechanisms.

To manage stress and PTSD among rescuers, various coping strategies can be employed. These include fostering a supportive work culture where open discussions about mental health are encouraged, implementing training programs focused on stress management techniques, and promoting regular physical exercise and mindfulness practices. Engaging in consistent physical activity can help reduce stress hormones and stimulate endorphins, which improve mood and energy levels. Mindfulness practices, like meditation and controlled breathing exercises, allow rescuers to focus their minds and reduce anxiety, enabling them to better handle high-pressure situations.

Critical incident debriefing and after-action reviews are essential tools in comprehending and improving rescue efforts. These processes involve structured reflections on incidents, allowing team members to analyze what happened, why it happened, and how future responses could be improved. Importantly, this approach is not limited to identifying mistakes but also recognizing successful strategies worth repeating. While emotional ventilation immediately following a traumatic event can sometimes be counterproductive, structured debriefings within a few days can facilitate learning and psychological recovery (Billings et al., 2023).

Support systems such as peer counseling and mental health resources play a pivotal role in maintaining the psychological health of rescuers. Peer support programs create avenues for individuals who have experienced similar traumatic events to share their experiences and provide mutual encouragement. These programs have shown positive results in reducing stigma associated with seeking help and increasing the likelihood of individuals utilizing available mental health services (Anderson et al., 2020). Furthermore, access to professional mental health resources ensures that rescuers dealing with more severe mental health issues receive appropriate treatment and support.

Analyzing case studies of both successful rescues and operational missteps provides valuable insights into improving future operations. By closely examining cases where outcomes were favorable, organizations can identify best practices and replicate these successes in different contexts. Simultaneously, studying instances where outcomes were less than ideal allows responders to learn from mistakes without casting blame, focusing instead on process improvement and risk mitigation.

Incorporating lessons learned from past experiences is instrumental in advancing the effectiveness of rescue operations. This continuous loop of action, reflection, and adaptation fosters a culture of learning and resilience within rescue teams. Such an approach not only enhances operational readiness but also contributes to individual and collective resilience by equipping rescuers with the knowledge and skills necessary to handle future challenges effectively.

It is important for organizations to institutionalize supportive measures to sustain the psychological resilience of their personnel. This includes ongoing education and training on mental health awareness and resilience-building techniques, as well as ensuring that leaders at all levels are equipped to recognize signs of mental distress among team members and know how to respond appropriately. By prioritizing mental health alongside physical preparedness, rescue organizations can better protect their personnel and optimize their operational capabilities.

Bringing It All Together

This chapter has provided a comprehensive exploration of rescue operations by focusing on essential principles, ethical considerations, and preparedness strategies. We have delved into the structured approaches that improve efficiency, drawing lessons from historical rescues like the Chilean mine incident and Hurricane Katrina. These examples highlighted the significance of international cooperation, technology integration, and the need for improved coordination in response efforts. By understanding these components, responders can enhance their ability to navigate complex situations effectively, prioritizing life safety and ensuring that all involved are protected.

Furthermore, we examined the critical balance between professionalism and empathy, emphasizing the importance of cultural sensitivity and ethical decision-making in high-pressure scenarios. The discussion on risk assessment and safety protocols underscored the necessity for continuous training and dynamic adaptation to changing environments. Personal Protective Equipment (PPE), Standard Operating Procedures (SOPs), and regular safety briefings form the backbone of operational readiness, ensuring responders are equipped to handle any situation confidently. Through scenario-based drills and technological advancements, teams enhance their preparedness while incorporating psychological resilience measures to protect mental health. Overall, this chapter aims to equip firefighters, first responders, and students with vital knowledge and insights to succeed in their demanding roles within emergency operations.

Chapter 1 Test Questions and Answers

What are the primary objectives of rescue operations?

Rescue operations aim to ensure life safety, provide medical care, and restore normalcy through effective resource allocation and coordination among emergency services.

How does a structured approach impact rescue operations?

It minimizes chaos and maximizes efficiency by establishing clear roles and responsibilities within participating agencies, ensuring rapid deployment and coordinated efforts.

What historical example demonstrated international cooperation in a rescue mission?

The 2010 Chilean mine rescue, where 33 miners were trapped underground for 69 days, showcased global collaboration and the use of advanced drilling techniques.

What lessons were learned from the Hurricane Katrina rescue efforts?

The response highlighted weaknesses in communication and resource distribution, leading to improvements in disaster management protocols.

How has technology influenced modern rescue techniques?

The use of drones, GPS, and sophisticated communication networks has enhanced search accuracy and response times, improving overall rescue efficiency.

Why is life safety the cornerstone of rescue operations?

It ensures the well-being of both victims and rescuers, requiring proper training and equipment to navigate hazardous situations effectively.

What role does scene control play in rescue operations?

Scene control manages hazards, secures the area, and establishes a command structure to ensure safe and efficient operations.

How does the Incident Command System (ICS) improve rescue operations?

ICS provides a standardized coordination approach, enabling efficient communication and resource allocation while integrating multiple agencies.

Why is informed consent an ethical duty in rescue operations?

It ensures victims understand the assistance provided and allows them to make decisions about their care, respecting their autonomy.

What is the importance of cultural sensitivity in rescue operations?

Recognizing and respecting diverse backgrounds helps build trust, facilitates cooperation, and improves overall rescue outcomes.

How can rescuers balance professionalism and empathy?

While adhering to protocols, rescuers must also provide compassionate support to victims, reinforcing the human-centered nature of their work.

What is the role of Personal Protective Equipment (PPE) in rescue operations?

PPE safeguards rescuers against physical, chemical, and environmental hazards, ensuring their safety during operations.

What are some examples of hazards in rescue operations?

Hazards include unstable structures, toxic gases, corrosive materials, inclement weather, and difficult terrain.

How do Standard Operating Procedures (SOPs) contribute to safety?

SOPs provide structured guidelines for executing tasks, ensuring consistency, efficiency, and risk reduction in rescue efforts.

Why are safety briefings and pre-rescue assessments essential?

They help outline objectives, communicate potential risks, and prepare responders for effective and coordinated action.

What is dynamic risk assessment, and why is it important?

It involves continuously evaluating risks and adapting strategies in real time, enhancing situational awareness and response effectiveness.

Why is physical and mental condition important for rescuers?

Rescuers operate under extreme conditions, requiring endurance, strength, agility, and mental resilience to handle high-stress environments.

What is the significance of certification in rescue operations?

Certification ensures rescuers are proficient in the latest life-saving techniques and adhere to operational standards.

How do scenario-based drills benefit rescue teams?

These drills simulate real-life emergencies, allowing responders to practice coordination, identify gaps, and refine tactical responses.

How does ongoing education contribute to effective rescue operations?

Continuous training keeps rescuers updated on new tools, methodologies, and strategies, improving their ability to adapt to evolving challenges.

Chapter 2: Assessing the Scene: Risk Analysis and Decision-Making Under Pressure

Assessing the scene is an essential part of ensuring effective and safe rescue operations. Performing a thorough assessment sets the stage for everything that follows, significantly impacting both the safety and success of the mission. Proper evaluation allows first responders to tailor their actions according to the unique challenges presented by each situation. This process involves not just recognizing obvious threats but also identifying less visible hazards that could compromise the safety of rescuers and victims alike. The skill and insight applied in these initial moments can dramatically alter the course of an operation, underscoring the importance of a well-conducted scene assessment.

This chapter delves into the strategies and methodologies for analyzing risks and making informed decisions under pressure. It explores the various steps involved in scene assessment, from initial observation to establishing command structures. Readers will gain insight into how dynamic risk analysis plays a key role in adapting to changing conditions during operations. Furthermore, the chapter examines decision-making processes under stress and the importance of clear communication among teams. By understanding these elements, readers will be better equipped to manage high-pressure situations, ensuring that they maintain safety while achieving their operational goals.

Introduction to Scene Assessment

Scene assessment is a pivotal aspect of effective rescue operations. It lays the groundwork for all subsequent actions and ensures that missions are conducted safely and efficiently. By performing thorough scene assessments, first responders can tailor their initial actions and responses appropriately, thus minimizing potential risks and enhancing the overall success of rescue efforts.

The foundation of any successful rescue operation begins with an accurate understanding of the environment in which the operation will take place. This involves identifying visible hazards, assessing structural integrity, and evaluating other environmental factors that could impact both rescuers and victims. For instance, in a collapsed structure scenario, trained Structures Specialists from Urban Search and Rescue teams play a crucial role in evaluating the security of the building before entry (<i>Structural Collapse Guide | Occupational Safety and Health Administration</i>, 2025). They determine how to mitigate structural hazards effectively, ensuring that rescue personnel can operate without unnecessary peril.

Inadequate scene assessment can have severe, sometimes catastrophic consequences. Failure to identify hidden dangers, such as exposed electrical wiring or unstable debris, increases the likelihood of accidents during operations, which can endanger lives and compromise the mission's success (Nowak, 2024). A poorly assessed site may result in unnecessary exposure to hazardous materials or conditions, resulting in operational failures or even fatalities.

Different incidents require unique approaches to scene assessment. For example, structural collapses demand careful evaluation of falling debris, potential explosive devices, and compromised utilities (<i>Structural Collapse Guide | Occupational Safety and Health Administration</i>, 2025). In contrast, scenes involving hazardous material spills necessitate different strategies, focusing on isolating contaminants and ensuring that responders utilize appropriate personal protective equipment.

Case studies from real-world scenarios emphasize the tangible impacts of successful or failed assessments on overall operations. Take, for instance, the tragic outcomes seen during rescue efforts following the Murrah Federal Office Building bombing in Oklahoma City and the terrorist attacks on September 11, 2001. Both events highlighted the importance of accurately assessing collapse zones and managing chaotic environments to prevent additional casualties and maintain operational integrity (<i>Structural Collapse Guide | Occupational Safety and Health Administration</i>, 2025).

Furthermore, lessons from these incidents stress the significance of not rushing into a scene unprepared. Initial surveys and the establishment of an Incident Command System are critical in organizing efforts and maintaining control over potentially volatile situations (Nowak, 2024). The calm, calculated approach allows for a comprehensive understanding of the risks involved and better communication among responding units, ensuring that everyone operates under a unified strategy.

Effective scene assessment also demands adaptability, as conditions can evolve rapidly. Responders must be prepared to adjust their tactics in response to new information or changes in the environment. Structural instabilities may occur unpredictably, requiring constant vigilance and reevaluation of the safety measures in place (Nowak, 2024). Adaptability is key to maintaining the safety of both rescuers and those they aim to save.

The Scene Assessment Process

Conducting a thorough scene assessment during rescue operations is a critical process that demands precision, attention to detail, and strategic planning. This section outlines steps in scene assessment, incorporating essential considerations for ensuring safety and operational efficiency.

The initial approach and size-up are fundamental components conducted from a safe distance to identify visible hazards and prioritize safety. As first responders arrive at the scene, the emphasis must be on maintaining their security while gaining an overview of the situation.

Before engaging with potential victims or hazards, rescuers should pause briefly to evaluate the surroundings, taking in details such as the layout of the area, weather conditions, and any ongoing threats like fires or unstable structures. This careful examination from a distance allows responders to form a strategic entry plan, minimizing unnecessary risks. (Drumm, 2022)

Intelligence gathering is the next crucial step in scene assessment. It involves evaluating the number of victims, their conditions, and identifying environmental dangers. Responders should swiftly but thoroughly assess the scene, noting any signs that indicate how many individuals are involved and any apparent injuries. This process includes looking for visual cues, such as abandoned vehicles or personal belongings that suggest victim presence. Understanding the environmental context—such as chemical spills or electrical hazards—is equally vital, as these factors can significantly affect both the rescue efforts and the safety of everyone involved. Gathering this information helps in prioritizing actions and allocating resources effectively. (2025)

Establishing incident command is integral to managing rescue operations efficiently. Designating roles and coordinating with responding units ensures streamlined efforts and clear communication. Once the initial assessment is complete, an incident command structure must be established promptly. This structure involves assigning specific responsibilities to personnel based on expertise and experience levels. The incident commander oversees overall operations, directing others to handle tasks such as medical attention, hazard management, and logistics. By forming this hierarchy, responders create a cohesive unit where each member knows their role, reducing confusion and enhancing operational effectiveness. Coordination among units, including police, EMS, and firefighters, is facilitated through structured communication and predefined action plans. (Drumm, 2022)

A key element of effective scene assessment is the communication of findings and assignments based on expertise to ensure efficient scene management. Clear, concise communication is paramount for aligning team efforts and avoiding misunderstandings. Information gathered during the assessment phase must be conveyed swiftly to all relevant parties. Radio communications and briefing sessions serve as platforms for sharing updates on victim status, identified hazards, and resource availability. Assignments should be delegated based on individual skills and capabilities, ensuring tasks are handled by those best equipped to manage them. This specialization not only enhances task execution but also maintains responder morale by aligning duties with expertise. (2025)

Risk Analysis: Identifying and Managing Hazards

In the realm of rescue operations, understanding and managing hazards effectively is crucial for first responders to ensure both their safety and that of civilians. These hazards can be broadly categorized into physical, chemical, biological, and environmental types, each presenting unique challenges that necessitate specific handling techniques.

Physical hazards, such as structural instability or debris in a building collapse scenario, require immediate and precise assessment. First responders must evaluate the potential risk of falling objects or unstable surfaces and implement measures like shoring or using protective gear to minimize these risks. Similarly, chemical hazards, which are especially prevalent in industrial accidents, demand careful identification and containment. For instance, exposure to hazardous substances like flammable gases may require evacuation procedures and specialized firefighting equipment designed to handle chemical fires (CCOHS, 2023).

Biological hazards, although less common, pose significant risks, particularly in settings like hospitals or during a pandemic. Responders need to be equipped with personal protective equipment (PPE) to prevent exposure to infectious agents. Understanding the nature of the biological threat allows for the appropriate deployment of decontamination protocols, ensuring the safety of all personnel involved (Down, 2024). Environmental hazards, including extreme weather conditions or rough terrain, also require adaptive strategies. Rescue teams may need to navigate through floods or work amidst intense temperatures, necessitating continuous monitoring and adjustment of tactics as conditions change.

Dynamic risk assessment (DRA) plays an integral role in managing these diverse hazards. Unlike traditional risk assessments, which may occur at the planning stage, DRA involves real-time evaluation and reevaluation of risks as new information becomes available. This approach requires responders to remain vigilant and agile, constantly updating their understanding of the situation to inform decision-making. By maintaining situational awareness and observing changes in the environment, responders can adapt their strategies quickly to mitigate emerging threats (CCOHS, 2023).

In situations where the risks are substantial, conducting a risk vs. reward evaluation becomes essential. This process involves weighing the inherent dangers against the potential benefits of taking action. If the risk to responders' lives outweighs the advantage of proceeding with a particular strategy, it might be safer to call in specialized resources, such as hazardous materials teams or structural engineers. For example, when dealing with a radiological incident, the expertise of individuals trained in radiation safety becomes invaluable. This ensures that any interaction with dangerous materials is conducted safely and efficiently, reducing the likelihood of harm (Down, 2024).

Continuous risk management underpins the decision-making processes within rescue operations, prioritizing both safety and operational objectives. Regular training, drills, and simulations help engrain these practices into the operational framework, preparing responders for high-pressure scenarios. These exercises are vital for developing the skills needed to conduct effective risk assessments and implement control measures swiftly. It also emphasizes the importance of teamwork and communication, ensuring that all team members understand their roles and responsibilities clearly (CCOHS, 2023).

Practical guidelines support dynamic risk assessment and risk vs. reward evaluation. A significant guideline for DRA involves establishing robust communication systems among team members. This facilitates the exchange of critical information about changing conditions and hazards, allowing for prompt updates to risk evaluations. An additional guideline for risk vs. reward evaluation entails setting predefined thresholds that indicate when it is necessary to escalate the response with additional resources or personnel. For instance, if a certain level of smoke or gas concentration is detected, the protocol could demand specialized intervention, ensuring an efficient and timely response (Down, 2024).

Decision-Making Under Pressure

In high-stress emergency situations, decision-making becomes a complex and dynamic process heavily influenced by stressors. Stress can impair cognitive functions, leading to errors in judgment which may jeopardize the success of rescue operations. It is critical for first responders to recognize how stress impacts their cognitive abilities and incorporate strategies that mitigate these effects.

Stress affects attention, perception, and memory retrieval—core components necessary for effective decision-making (Penney et al., 2022). For instance, during a fire incident, a firefighter may become so focused on one task that they overlook emerging hazards or fail to communicate crucial information. In such scenarios, errors like tunnel vision—a narrowed focus hindering awareness of the broader scene—can occur. Therefore, training programs should address common cognitive biases and teach responders techniques to maintain situational awareness under pressure.

Utilizing decision-making models such as the OODA Loop (Observe, Orient, Decide, Act) provides a structured approach that helps reduce complexities involved in rapid assessments. This model emphasizes continuous observation and orientation within rapidly changing environments allowing responders to adapt quickly (Cohen-Hatton et al., 2015; Okoli et al., 2016). By following a deliberate sequence of steps, responders enhance their ability to make informed decisions swiftly, turning potential confusion into actionable insights.

Structured checklists further aid in organizing thought processes by ensuring all critical aspects are considered systematically. These lists act as mental reminders, helping responders cover essential elements and reduce the likelihood of oversight amid chaos. Structured decision-making methods equip responders with predefined procedures that balance speed and accuracy, facilitating decisive action in unpredictable circumstances (Perona et al., 2019).

Risk-based decision-making prioritizes safety and operational effectiveness by considering both immediate threats and the potential benefits of actions taken. Decision-makers assess whether conditions warrant immediate intervention or if additional data collection would optimize results without compromising safety (Bak et al., 2003). For example, in an urban search-and-rescue mission, team leaders weigh the risks of entering an unstable building against the urgency of possible survivors trapped inside. Calculated actions, based on comprehensive risk assessment, lead to successful outcomes while minimizing harm.

Recognizing when to take immediate action versus engaging in methodical information gathering is crucial during emergencies. Some scenarios demand quick reflexes due to imminent danger posed to life or environment. Conversely, situations where thorough analysis reveals unforeseen risks necessitate patience and strategic planning. A balanced approach empowers responders to navigate the fine line between impulsiveness and over-cautiousness (Klein et al., 2010).

Integrating situational cues with past experiences enhances decision-making efficacy. Responders develop mental models from accumulated knowledge through training and field experience, allowing them to predict potential developments accurately. Fast-paced situations often require intuitive judgments relying on these mental shortcuts, particularly in familiar scenarios (Kahneman and Klein, 2009). However, integrating intuition with analytical reasoning ensures adaptability to unfamiliar challenges, reducing reliance on assumptions that could prove detrimental.

Training programs designed for first responders must emphasize the importance of flexibility and adaptability within their decision-making frameworks. Developing adaptive expertise enables individuals to modify established routines according to evolving situations, making them better equipped to handle unexpected challenges (Seiler et al., 2010). Continuous training fosters agility, preparing personnel to execute precise decisions regardless of external stressors.

Institutional support plays a vital role in reinforcing effective decision-making practices. Organizations should encourage cross-disciplinary collaboration and communication, promoting shared learning among different teams involved in emergency response (Penney et al., 2022). Furthermore, ongoing evaluation and refinement of decision-making protocols contribute to improved outcomes and create a culture that values innovation.

Regular debriefing sessions provide opportunities for responders to reflect on their decision-making processes, learn from successes and failures, and identify areas for improvement. Debriefs foster an environment of openness, encouraging constructive feedback that strengthens the collective ability to perform under pressure. By continuously refining their approaches, responders increase resilience against stress-induced errors.

Effective Communication and Tactical Adjustments

In any rescue operation, communication and adaptability play crucial roles in ensuring the safety and efficiency of all involved. Clear radio communication is foundational to successful emergency response operations. When responders utilize standardized terms, it minimizes misunderstandings and ensures that every member of the team understands the situation. The use of common terminology and protocols is vital, particularly when dealing with high-pressure situations where every second counts.

Effective coordination among multiple agencies, such as fire services, Emergency Medical Services (EMS), and police departments, also necessitates clear communication channels. During complex emergencies, different agencies must work together seamlessly to achieve a unified operational goal. For instance, integrating efforts between firefighters and EMS can mean the difference between life and death for victims in need of urgent medical attention while managing a hazardous environment simultaneously. By aligning strategies and objectives through open communication lines, each agency can utilize its expertise effectively without overlap or missing critical elements in the response plan (Staff, 2022).

Adaptive strategies are essential in constantly changing scenarios, such as structural shifts during rescue operations. Natural disasters or accidents can lead to unexpected environmental changes, requiring teams to reassess their plans and redistribute resources swiftly. For example, if a building's structural integrity suddenly becomes compromised, responders must quickly adapt by deploying additional support personnel or tools to ensure ongoing safety for both rescuers and victims. Regular training and simulation exercises prepare emergency teams to think on their feet and respond dynamically, maintaining safety and efficiency under pressure.

Furthermore, emergency withdrawal procedures play a significant role in protecting personnel during life-threatening situations. These procedures emphasize the importance of accountability and strategic retreat when necessary. It is crucial that all team members understand when and how to withdraw from a scene to prevent further risk to themselves and others. A well-documented incident command system helps maintain an accurate headcount and track the movements of all personnel, thereby reducing the likelihood of individuals being left behind or injured during evacuation processes. This structured approach allows for controlled withdrawal, prioritizing the safety of all responders involved.

To enhance these operations, implementing a coordinated communication infrastructure, like WebEOC, can streamline inter-agency data sharing and decision-making processes. When emergency management departments integrate such systems, they enable real-time information dissemination across various agencies, allowing them to act efficiently and effectively based on evolving field conditions (Staff, 2022). In a similar vein, crisis communication strategies should be tailored to diverse communities to ensure accessibility and clarity. Emergency managers must consider language barriers, technological access, and cultural differences when crafting messages to ensure that communication is both effective and inclusive (<i>Crisis Communication Strategies for Emergency and Crisis Management Officials</i>, 2021).

Ultimately, to execute successful rescue operations amid evolving scenes, all involved parties need to foster a culture of continuous learning and improvement. Regular debriefings after incidents allow agencies to evaluate their performance, identify areas for enhancement, and adjust their guidelines and protocols accordingly. Through collaborative efforts, first responders, and emergency managers can refine their approaches to communication and adaptability, benefiting from shared experiences and collective knowledge.

Concluding Thoughts

This chapter has examined the essential strategies and processes involved in scene assessment, risk analysis, and decision-making during rescue operations. We've explored how thorough initial evaluations can set the stage for effective responses, emphasizing the need to recognize visible hazards while maintaining a focus on safety. By identifying potential dangers early on, first responders can develop informed action plans, ensuring that efforts are both strategic and safe. The exploration of different incident types underscored the importance of adapting assessment techniques and tailoring them to specific situations, thus improving operational outcomes.

Understanding and managing risks effectively remain at the heart of successful rescue missions. We have discussed the significance of dynamic risk assessment and how it empowers responders to adapt quickly as conditions evolve. The integration of structured decision-making models emphasizes the balance between speed and accuracy under stress, reinforcing the necessity for clear communication and flexibility in tactics. By fostering an environment of continuous learning and adaptation, first responders are better equipped to protect themselves and those they serve, ensuring the overall success of their operations.

Chapter 2 Test Questions and Answers

What is scene assessment in rescue operations?

Scene assessment is the process of evaluating the environment where a rescue operation is taking place to identify hazards, risks, and necessary safety precautions. It sets the foundation for safe and effective rescue efforts.

Why is scene assessment crucial for first responders?

Scene assessment is crucial because it helps responders identify visible and hidden hazards, enabling them to make informed decisions that ensure the safety of both victims and rescuers.

What should responders prioritize during initial scene assessment?

Responders should prioritize safety, identify visible hazards, assess the structural integrity of buildings, and evaluate environmental conditions such as weather or fires.

How can inadequate scene assessment affect rescue operations?

Inadequate scene assessment can lead to overlooked dangers such as exposed electrical wiring or unstable structures, which could result in accidents, injury, or failure of the rescue operation.

What is the role of Structures Specialists in a collapsed structure scenario?

Structures Specialists assess the stability of the collapsed building, identifying risks like falling debris or compromised utilities and ensuring that rescue teams can safely enter and operate within the structure.

How do hazards in different incidents affect scene assessment strategies?

Each incident presents unique challenges. For example, in structural collapses, the focus is on debris and building integrity, while in hazardous material spills, the focus is on containment and personal protective equipment (PPE).

Why were the Murrah Federal Office Building bombing and September 11 attacks significant in terms of scene assessment?

These events emphasized the importance of accurately assessing collapse zones, managing chaotic environments, and ensuring safety to avoid further casualties and maintain operational integrity.

What is the Incident Command System (ICS), and why is it important?

The ICS is a structured approach to managing rescue operations by assigning roles and responsibilities. It helps streamline efforts, coordinate teams, and ensure clear communication across agencies.

How should first responders approach the scene initially?

First responders should approach the scene from a safe distance, conduct a size-up to identify visible hazards, and gather intelligence before engaging them with victims or hazards.

What is intelligence gathering in scene assessment?

Intelligence gathering involves assessing the number of victims, their conditions, and identifying environmental dangers. It helps responders prioritize actions and allocate resources efficiently.

Why is clear communication vital during scene assessment?

Clear communication ensures that all responders are informed about identified hazards, the status of victims, and available resources, reducing the likelihood of misunderstandings and mistakes.

What types of hazards should first responders assess during scene evaluation?

Responders should assess physical, chemical, biological, and environmental hazards, each requiring specific strategies for management and mitigation.

What is dynamic risk analysis (DRA)?

DRA is the continuous, real-time evaluation of risks as new information becomes available. It allows responders to adjust strategies and maintain situational awareness during operations.

What is the importance of risk vs. reward evaluation in rescue operations?

Risk vs. reward evaluation involves weighing the potential benefits of an action against the dangers it poses to rescuers. It helps determine when to proceed or when to call for specialized resources.

How does stress affect decision-making in high-pressure situations?

Stress can impair cognitive functions such as attention, memory, and perception, potentially leading to errors in judgment. It is crucial for responders to recognize stress's impact and use strategies to mitigate its effects.

What is the OODA Loop, and how does it assist in decision-making under pressure?

The OODA Loop (Observe, Orient, Decide, Act) is a decision-making model that helps responders assess rapidly changing situations and make informed decisions quickly and effectively.

How do structured checklists help responders during rescue operations?

Structured checklists help organize thought processes, ensuring that all critical aspects of the operation are addressed systematically and reducing the likelihood of oversight.

What is risk-based decision-making in rescue operations?

Risk-based decision-making involves assessing immediate threats and potential benefits before taking action, ensuring that decisions prioritize safety while achieving operational objectives.

Why is situational awareness important for decision-making in emergencies?

Situational awareness helps responders assess the current conditions, predict potential developments, and adjust their strategies accordingly to minimize risks and ensure effective response.

How does adaptability influence rescue operations?

Adaptability allows responders to adjust to evolving situations, such as changes in structural integrity or environmental conditions, ensuring that safety is maintained, and operations continue smoothly.

Chapter 3: Technical Rescue: Tools, Techniques, and Tactical Approaches

Technical rescue operations are a vital aspect of emergency response, distinguishing themselves from traditional firefighting and emergency medical services. These operations involve unique scenarios that require specialized training, skills, and equipment distinct from those used in more common emergency situations. Firefighters and first responders engaged in technical rescues must be adept in various areas, including confined space entry, high-angle rope work, and structural stabilization. The specialized nature of these rescues demands not only a deep understanding of the physical and technological aspects involved but also a mastery of tactical approaches necessary for effective and safe intervention during complex incidents.

This chapter delves into the essential components that form the backbone of technical rescue operations. Readers will gain insights into the variety of tools utilized, along with techniques that are critical for responding to diverse rescue scenarios. The chapter will also explore the tactical approaches that enable teams to execute their missions efficiently while prioritizing safety. Through this exploration, readers will understand the specialized training required to navigate these demanding environments effectively. They will also learn about the evolving nature of technical rescue as it adapts to advancements in technology and changes in best practices, ensuring readiness for any challenge that may arise on the field.

Introduction to Technical Rescue

Differentiating technical rescue from traditional emergency services is crucial for understanding the complexities of emergency management. While conventional firefighting and emergency medical services (EMS) are well-established and essential components of public safety, technical rescue introduces a specialized dimension that demands distinctive training and tools. This distinction is vital in coordinating efficient rescue operations during highly complex incidents.

Technical rescues are set apart by their requirement for advanced training and specialized equipment. Unlike routine firefighting or EMS tasks, these rescues often involve scenarios where standard protocols are insufficient. For instance, firefighters might handle structural fires or car accidents regularly, but technical rescue calls demand expertise in unique areas such as confined space entry, high-angle rope work, or trench stabilization. These scenarios require precise knowledge and skills beyond the scope of typical emergency service training.

The need for specialized training becomes evident when considering the diverse types of technical rescues. Confined space rescues, for example, involve entering perilous environments with limited access, requiring rescuers to operate under conditions of potential oxygen deficiency or toxic gases. Rope rescues, on the other hand, involve high-angle operations often necessary for reaching victims trapped on cliffs or towers. Each type of technical rescue poses distinct hazards and challenges, necessitating comprehensive, scenario-specific training and preparation.

The history of technical rescue has played a significant role in shaping modern standards and best practices. Historical events have underscored the necessity of equipping responders with the right skills and tools for highly specialized situations. Notable cases, like the 1980 Calgary Tower rescue, have highlighted the role of technical rescue in life-saving efforts and led to the development of more rigorous training standards and equipment innovations. Such incidents not only demonstrate the importance of being prepared for rare, yet critical emergencies, but also illustrate how past experiences inform current rescue protocols and procedures.

Understanding the definition and importance of technical rescue is essential for recognizing its vital role in emergency management. Technical rescues address emergencies that surpass the capabilities of standard response methods, serving as critical components of comprehensive emergency preparedness strategies. These operations provide the means to take action in complex and potentially life-threatening situations, thus reinforcing the overall efficacy of an emergency management system.

Technical rescues play an invaluable part in expanding the reach and effectiveness of public safety responses. They ensure that no matter how challenging the circumstances, trained personnel are ready to face the demands with precision and expertise. This specialization empowers first responders to tackle extraordinary emergencies safely and efficiently, minimizing risks to both victims and crew members involved.

As technology continues to advance, the technical rescue field evolves with it, integrating new tools and techniques into practice. The development of improved equipment, such as automated backup devices and versatile rope systems, enhances the safety and speed of rescue operations. Innovations like these are born from a combination of research, experience, and historical lessons learned, all aimed at advancing the effectiveness and safety of technical rescue missions.

Fundamental Principles of Technical Rescue

In technical rescue operations, decision-making hinges upon a rigorous risk versus benefit analysis. This principle demands evaluating the potential risks to rescuers and victims against the probability of successful outcomes. When faced with a high-risk scenario, such as entering a structurally compromised building or conducting a swift water rescue, first responders must assess whether the benefits of action justify the risks involved. This decision-making process relies on data and expertise to ensure that all factors are considered, from environmental hazards to equipment limitations. By prioritizing this analysis, rescue teams can determine if the operation is viable, ensuring the safety of both the team and those they aim to save.

The adoption of the Incident Command System (ICS) is crucial for effective tactical coordination during technical rescues. The ICS organizes resources and establishes clear roles within emergency response teams, allowing for streamlined communication and better decision-making. For instance, the system designates specific positions like the Incident Commander, who oversees and manages the overall response strategy, and the Safety Officer, responsible for monitoring onsite safety conditions. These well-defined roles enhance operational efficiency and mitigate confusion during high-pressure situations, ultimately increasing the likelihood of a successful rescue. The establishment of distinct command structures ensures all team members understand their roles and responsibilities, which is essential in maintaining order and focus during complex rescue operations.

An important aspect of technical rescue is the procedural framework guiding each operation, commonly referred to as rescue phases. These phases begin with a thorough hazard assessment, where rescuers evaluate potential threats, such as structural instability or hazardous materials, that could impact the operation. Following the assessment, stabilization efforts are made to ensure the environment does not pose additional dangers. This could involve shoring up weakened structures or securing a safe perimeter. Once the area is deemed stable, the priority shifts to victim extrication, applying techniques and tools appropriate for safely removing individuals from perilous situations. After completing the physical rescue, a post-operation debriefing takes place, where the team reviews actions taken to identify lessons learned and areas for improvement. This debriefing is critical for enhancing future mission success and safety protocols.

These core principles—risk versus benefit analysis, strategic use of the ICS, and structured rescue phases—are instrumental in minimizing risks and improving the success rate of technical rescue operations. When executed effectively, they facilitate a more controlled and predictable environment, allowing rescuers to focus on their primary objective: saving lives. Moreover, consistently adhering to these principles fosters a culture of preparedness and continuous learning among first responders, equipping them with the knowledge and skills necessary to handle diverse and dynamic challenges they may encounter on the field.

To illustrate these concepts, consider a scenario where first responders are called to a collapsed building following an earthquake. Initial risk assessments reveal unstable debris and gas leaks, presenting significant hazards. Utilizing the ICS, the Incident Commander assesses resources and assigns roles—Rescue Group Supervisor, Medical Team Lead, and Safety Officer—to ensure coordinated efforts. The team conducts a hazard assessment before stabilizing the site, using shoring timbers to prevent further collapse. During extrication, rescuers use specialized equipment to reach trapped victims safely, while medical personnel prepare to administer immediate care. After successfully rescuing individuals, a detailed debrief analyzes the operation's strengths and shortcomings, reinforcing training protocols and strategies for future incidents.

Tools and Equipment for Technical Rescues

Technical rescue operations require a meticulous understanding of the tools and equipment essential for ensuring safety and effectiveness. The foundation of any successful rescue mission lies in the appropriate use and maintenance of these tools, which vary from personal protective equipment to specialized machinery designed for specific scenarios.

Personal Protective Equipment (PPE) serves as the first line of defense for rescuers in diverse situations. Helmets, gloves, and specialty gear are vital for protecting responders against environmental hazards, debris, and other unforeseen dangers. For instance, helmets must meet safety standards to withstand impacts, while gloves should provide both dexterity and protection against cuts and abrasions. In high-risk environments, such as chemical spill sites or fire zones, additional gear like respirators or flame-resistant suits may be required. Understanding and choosing the right PPE is crucial; it not only ensures personal safety but also influences the overall success of the operation by allowing rescuers to perform their tasks efficiently and confidently.

Beyond PPE, technical rescues utilize an array of basic and advanced tools tailored to specific rescue scenarios. Ropes and pulleys are fundamental in many operations, facilitating the safe movement of individuals and equipment across difficult terrains. These simple yet effective tools can be configured into complex systems that support high-angle rescues or provide mechanical advantage in lifting heavy loads. Hydraulic tools come into play when precision and power are necessary, such as during vehicle extrications. These tools allow rescuers to cut through metal swiftly and safely, freeing trapped individuals with minimal risk. Similarly, thermal imaging cameras represent a significant technological advancement, enabling rescuers to detect heat signatures through smoke or darkness, revealing hidden threats or locating individuals in distress. The integration of such technology enhances situational awareness and decision-making, providing a strategic advantage during critical moments.

While having the right tools is imperative, regular maintenance and inspection are equally crucial to ensure readiness and compliance with safety standards. Equipment should be inspected periodically according to guidelines set forth by manufacturers and safety authorities. Regular checks help identify wear and tear, potential malfunctions, or other issues that could compromise safety during an operation (Ford, 2023). Additionally, documentation and communication are integral to this process; maintaining detailed records of inspections and repairs ensures traceability and accountability within teams (Tabibian, 2025). This organized approach supports proactive maintenance practices, reducing the likelihood of unexpected equipment failures that could jeopardize a rescue mission.

Proper training in the use of equipment is paramount for executing effective and safe rescue operations. Rescue personnel must be well-versed not only in operating various tools but also in understanding their limitations and potential risks. Training drills and exercises simulate real-world scenarios, allowing teams to hone their skills and improve coordination under controlled conditions. Such experiences foster confidence and competence, preparing them for the unpredictable nature of actual emergencies. Furthermore, ongoing education and skill refinement enable rescuers to stay updated on new technologies and methodologies, ensuring that their knowledge evolves alongside advancements in safety equipment.

Rope Rescue Operations

In the realm of technical rescue, rope operations are a specialized domain requiring both skill and precision. To understand the nuances involved, it's crucial to differentiate between high-angle and low-angle rescues. High-angle rescues refer to situations where rescuers operate on steep inclines or vertical spaces exceeding 60 degrees. In these scenarios, the reliance on ropes and associated gear is paramount for the safety of both patients and rescuers. On the other hand, low-angle rescues take place on slopes less than 60 degrees, where the terrain provides more support but still requires the use of ropes for effective maneuvering. Understanding these differences is vital in planning and executing operations as it influences equipment selection, the number of personnel required, and the specific techniques employed (Dunn, 2024).

A foundational aspect of successful rope rescue operations is proficiency in knot tying and understanding rope systems like belay and haul systems. Knots form the backbone of any rescue setup, providing security and stability to the entire operation. Rescuers must master a variety of knots such as figure eight, bowline, and clove hitch, each serving distinct purposes in different scenarios. The belay system, meanwhile, is essential for safeguarding against accidental falls. It acts as a backup line, ensuring that even if the main line fails, the system will prevent further descent. Similarly, haul systems are critical when lifting victims or heavy loads. By employing mechanical advantage through pulleys, these systems significantly reduce the physical effort required, facilitating smoother and safer rescues (TheDoctorFeelGood, 2025).

Rope rescue techniques aren't just about securing equipment; they encompass the methods used by rescuers to navigate challenging environments. Techniques for descending and ascending, such as rappelling, are integral to these operations. Rappelling allows rescuers to descend heights with controlled speed and positioning, which is crucial in reaching victims positioned at various vertical points. Ascending, often achieved using mechanical ascenders or prusik knots, enables rescuers to climb back up to secure locations once they've reached the victim. Mastery of these techniques boosts the effectiveness and efficiency of rescue teams, allowing them to adapt quickly to the ever-changing dynamics of real-world rescue scenarios (TheDoctorFeelGood, 2025).

An equally important component of rope rescue is victim packaging and evacuation strategies. Proper victim packaging is essential to minimize further injuries during transportation. This involves securely immobilizing the patient using stretchers or specialized spinal boards to ensure safety while moving through complex terrains. Effective packaging not only protects the victim but also simplifies the task for rescuers by keeping the load stable and easier to control. Evacuation strategies might involve horizontal or vertical lifts depending on the situation. In high-angle rescues, vertical lifts using harnesses and stretchers are more common, whereas low-angle rescues may allow for more straightforward horizontal evacuations.

Implementing guidelines for descending and ascending techniques enhances the safety and efficacy of operations. When rescuers utilize rappelling, they must be trained in managing their speed and maintaining control over their body position to avoid accidents. Guided training sessions focusing on these skills can build confidence and competence among rescuers. Similarly, the principles of victim packaging and evacuation necessitate guidelines to ensure consistency and safety across all operations. These include steps for assessing patient condition, selecting appropriate packaging methods, and continuously monitoring victim status during transport to adapt strategies as needed.

By equipping themselves with an in-depth knowledge of rope systems, including belay and haul systems, rescuers assure operational success. These systems enable teams to navigate hazardous environments while minimizing risks effectively. Proficiency in knot tying remains critical, as it forms the basis upon which all other systems rely. Through practice and commitment to mastering these elements, rescue professionals can ensure their readiness for diverse emergency scenarios. Additionally, constant evolution in equipment and techniques demands continuous learning and adaptability from those involved in rope rescue missions.

Confined Space Rescue

Confined space rescues represent some of the most complex and hazardous operations first responders may face. They demand a nuanced understanding of the potential dangers and regulatory requirements that ensure the safety of both rescuers and victims. Mastery of these essential elements is crucial for executing successful confined space rescues.

The inherent hazards of confined spaces, such as oxygen deficiency and toxic gases, must be thoroughly understood before any rescue operation can commence. Confined spaces are often poorly ventilated, leading to an accumulation of harmful substances that can pose life-threatening risks. Oxygen levels in these environments can deplete due to various factors including chemical reactions or biological processes, leaving insufficient oxygen for safe breathing. According to OSHA guidelines, maintaining oxygen levels between 19.5% and 23.5% is crucial for safe entry into these spaces (<i>Working in Confined Spaces: Safety Guide [IAuditor]</i>, 2018). Toxic gases like carbon monoxide and hydrogen sulfide are common culprits that lead to significant health risks, potentially causing unconsciousness or even death with prolonged exposure. Therefore, understanding the specific atmosphere within a confined space before entry cannot be overstated.

Complying with OSHA regulations is imperative not only for legal reasons but also for ensuring the well-being of all involved. These regulations highlight the necessity of evaluating spaces to determine if they require a permit, which would indicate substantial potential risks and necessitate strict entry and safety procedures (<i>Working in Confined Spaces: Safety Guide [IAuditor]</i>, 2018). The designation of a permit-required confined space often means that critical measures, such as proper signage and entry protocols, must be followed. This compliance ensures that everyone on-site recognizes the potential hazards and understands the procedures necessary to mitigate them.

A key component of managing air quality and ensuring safety is effective ventilation and atmospheric monitoring. Atmospheric testing should always occur before anyone enters the confined space, using devices capable of detecting oxygen levels, flammable gases, and toxic substances such as carbon monoxide and hydrogen sulfide (Gislason, 2024). Continual monitoring is vital throughout the duration of the operation, as conditions within the space can change rapidly.

Mechanical ventilation systems play a significant role in improving air quality by introducing fresh air and expelling contaminated air. This process significantly reduces the risk of exposure to dangerous gases and improves the overall safety for rescue teams entering the space.

Specialized rescue techniques are essential tools in the arsenal of any technical rescuer dealing with confined spaces. One of the most commonly used methods involves tripod systems, which allow safe and efficient entry and extraction from vertical confined spaces. These systems provide a stable anchor point for ropes and pulleys, facilitating controlled descents and ascents. Equally important are supplied-air respirators, which provide a steady supply of breathable air to rescuers operating in environments where atmospheric contaminants cannot be adequately removed through ventilation alone. These advanced systems are indispensable, particularly when the environment poses a persistent threat despite other measures taken.

Training responders in these specialized techniques is another critical aspect of confined space rescue preparedness. Training programs should include regular hands-on practice with equipment like tripod systems and respirators to ensure proficiency. Practicing various rescue scenarios prepares teams for real-life situations, enabling them to respond swiftly and effectively. Moreover, it's essential to have personnel trained to perform atmospheric monitoring, as their expertise in recognizing and reacting to changes in air quality can mean the difference between a successful rescue and a tragic outcome.

Summary and Reflections

This chapter has delved into the multifaceted world of technical rescue operations, emphasizing the specialized skills and equipment that distinguish these efforts from more conventional emergency responses. We've explored the role of advanced training and tools required to tackle unique challenges in scenarios like confined spaces and high-angle rescues. Historical cases have demonstrated the importance of being prepared for rare, yet critical emergencies, underscoring the need for responders to be well-equipped with the right procedures and knowledge. The text also covered the significance of coordinating resources effectively through systems like the Incident Command System, which enhances operational efficiency and improves outcomes.

The discussion extended to examining essential tools and techniques crucial for technical rescue operations. From personal protective equipment ensuring safety in diverse environments to the strategic use of ropes, pulleys, and hydraulic tools tailored for specific situations, we highlighted how these components contribute to effective rescue missions. Emphasizing ongoing training and adherence to safety standards ensures teams are ready to face dynamic challenges on the field. As advances in technology continue to shape rescue practices, keeping abreast of new developments allows first responders to execute their duties more safely and efficiently.

Chapter 3 Test Questions and Answers

What differentiates technical rescue from traditional emergency services?

Technical rescue requires specialized training and tools to handle complex scenarios such as confined space entry, high-angle rope work, or trench stabilization, which are beyond the scope of standard firefighting or EMS.

What types of technical rescues require advanced knowledge and skills?

Confined space rescues, rope rescues, and trench rescues require advanced knowledge and skills due to their unique challenges and hazards.

Why is specialized training necessary for technical rescues?

Specialized training is necessary because technical rescues involve complex environments and hazards that standard firefighting or EMS protocols cannot address, such as toxic gases, structural instability, and confined spaces.

Can traditional firefighting or EMS methods handle all emergencies?

No, technical rescues address emergencies that surpass the capabilities of standard methods, requiring specialized techniques and equipment.

What are confined space rescues, and why are they particularly dangerous?

Confined space rescues involve rescuing individuals from hazardous environments with limited access. They are dangerous due to the potential for oxygen deficiency, toxic gases, and physical constraints.

How does rope rescue differ from traditional rescue methods?

Rope rescues require specialized equipment and techniques, including high-angle and low-angle operations, to safely extricate victims from difficult or steep terrain, which is not typically addressed by traditional methods.

What historical event significantly influenced modern technical rescue practices?

The 1980 Calgary Tower rescue highlighted the need for technical rescue and led to the development of more rigorous training standards and equipment innovations.

What is the primary decision-making principle in technical rescue?

The primary decision-making principle is a risk versus benefit analysis, where rescuers evaluate the potential risks to themselves and victims against the likelihood of a successful outcome.

What role does the Incident Command System (ICS) play in technical rescues?

ICS provides a structured command and coordination system that ensures clear roles and responsibilities, facilitating efficient decision-making and communication during complex technical rescues.

What are the key phases in a technical rescue operation?

The key phases include hazard assessment, stabilization, victim extrication, and post-operation debriefing.

How does regular equipment maintenance contribute to successful rescue operations?

Regular maintenance ensures that equipment is ready for use, reduces the risk of malfunction, and complies with safety standards, which are crucial during rescue operations.

Why is Personal Protective Equipment (PPE) critical in technical rescues?

PPE protects rescuers from environmental hazards, debris, and other dangers during rescue operations, ensuring their safety and ability to perform tasks effectively.

What is the function of ropes and pulleys in technical rescues?

Ropes and pulleys are essential for high-angle rescues and lifting heavy loads, providing mechanical advantage and supporting safe movement of victims and rescuers.

What are the risks involved in confined space rescues?

The risks include oxygen deficiency, exposure to toxic gases, and physical injury due to limited access and difficult working conditions.

How are rope rescue operations classified based on terrain?

Rope rescue operations are classified as high-angle rescues (steep inclines or vertical spaces over 60 degrees) and low-angle rescues (slopes less than 60 degrees).

What is the importance of knot tying in rope rescue operations?

Knot tying is critical for ensuring the safety and stability of rescue systems. Different knots are used for various purposes, providing security for rescuers and victims.

What specialized equipment is used in confined space rescues?

Specialized equipment includes tripod systems for safe entry and extraction, supplied-air respirators for breathable air, and atmospheric monitoring devices to assess air quality.

Why is atmospheric monitoring vital in confined space rescues?

Atmospheric monitoring is essential to detect harmful gases, oxygen levels, and other atmospheric hazards that could pose serious risks to both rescuers and victims.

What is the role of a debriefing after a technical rescue operation?

The debriefing analyzes the operation's strengths and weaknesses, allowing responders to learn from the experience and improve future rescue protocols and safety measures.

How has technology impacted technical rescue operations?

Technology has introduced advanced equipment such as thermal imaging cameras, automated backup devices, and versatile rope systems, improving the speed, safety, and effectiveness of rescue operations.

Chapter 4: Water Rescues: Swiftwater, Floods, and Ice Emergencies

Water rescues demand a highly specialized skill set, given the various environments in which they occur—swift waters, floods, and icy terrains all present unique and formidable challenges to first responders. Recognizing the diversity of these emergencies is crucial as each situation requires a distinct approach, keen awareness, and strategic thinking. The perilous nature of swift currents, overwhelming floodwaters, and deceptively fragile ice surfaces necessitates not only physical preparedness but also an understanding of environmental dynamics and potential hazards. First responders must be equipped with the right tools and techniques to effectively manage these complex scenarios while ensuring the safety of both rescuers and those in distress.

In this chapter, we delve into the fundamental principles and tactics essential for executing successful water rescues across diverse environments. Readers will gain insights into the specific challenges posed by swift waters, such as the rapid development of dangerous conditions and the presence of submerged obstacles. We further explore the unpredictability and extensive impact often associated with floods, emphasizing the importance of adaptability and situational awareness. Ice rescues are examined through the lens of safety and precaution, highlighting the need for specialized equipment and training. Throughout the chapter, the focus remains on fostering a comprehensive understanding of how to navigate these risky situations effectively, prioritizing preparedness, and leveraging lessons learned from past incidents to enhance rescue outcomes.

Introduction to Water Rescues

In the realm of emergency response, water rescues are among the most demanding operations due to their inherent complexity and danger. The diversity in water environments necessitates a comprehensive understanding of the specific dynamics and hazards present in each scenario. Whether it is navigating the treacherous currents of swift waters, managing the overwhelming force of floodwaters, or confronting the deceptive stability of ice-covered surfaces, each rescue situation presents unique challenges that demand tailored strategies and advanced training.

Differentiating between swiftwater, flood, and ice rescues forms the cornerstone of effective preparation and response. Swiftwater rescues involve operating in fast-moving water environments like rivers or rapids, where the speed of the water can quickly turn a stable situation into a volatile one. The dynamics at play require rescuers to have a keen understanding of hydrology and the ability to assess hazards such as strainers—natural or man-made obstacles that can trap individuals—or hydraulic jumps, which can forcefully submerge anyone caught within them (Hurst, 2024). Flood rescues, on the other hand, are often characterized by extensive and rapidly changing conditions, sometimes covering large urban areas. These situations can become chaotic, with strong currents carrying debris that poses additional dangers to both rescuers and victims. Ice rescues present a different sort of challenge; here, the seemingly solid surface can fracture unexpectedly, making it perilous for responders to traverse.

Historical cases provide invaluable lessons that shape current practices in water rescue operations. The 2018 Hurricane Florence incident serves as an essential case study, highlighting the necessity of deploying specialized teams such as the Type II swiftwater rescue units from Massachusetts. These teams were tasked not only with physical rescues but also with logistical coordination across vast areas impacted by flooding (McCabe, 2019). Another significant incident was the Winter Storm Riley, where technical rescue teams had to adapt to fluctuating temperatures and varying ice thicknesses while maintaining high levels of thermal protection and buoyancy. These historical events underscore the importance of adaptability and preemptive planning in rescue missions.

Water rescue environments are unpredictable and specialized, each demanding a distinct approach. The ever-changing nature of water bodies means that what is viable in one moment may not be safe in the next. For instance, urban floods can suddenly intensify due to factors like breached levees or broken storm drains. Similarly, ice conditions can shift from stable to dangerous due to temperature fluctuations and movements beneath the surface. This unpredictability requires rescuers to conduct continuous risk assessments and adapt their strategies accordingly (Hurst, 2024).

The specialized nature of these operations cannot be understated. Each type of water rescue demands not only physical fitness and mental resilience but also a mastery of technical skills specific to the environment at hand. For example, swiftwater rescues might necessitate the use of rope systems and inflatable devices, techniques acquired through rigorous training programs that adhere to standards set by the National Fire Protection Association (NFPA) (Hurst, 2024). The deft handling of equipment, along with the ability to perform under pressure, is crucial in any successful operation.

Training plays a pivotal role in preparing responders for the diverse scenarios that water rescues entail. The harsh reality is that these emergencies are some of the most dangerous that first responders will face. Rigorous and continual training ensures that personnel are equipped with the necessary skills to protect themselves and others. Programs specifically designed for swiftwater or ice rescues teach responders how to employ safety gear, execute effective self-rescue techniques, and stabilize victims before extrication (McCabe, 2019). Such comprehensive instruction is vital because the margin for error is slim in these life-threatening environments. The collective knowledge gained through regular drills, simulations, and the exchange of best practices can significantly reduce the risks involved and enhance the likelihood of a successful rescue outcome.

Hydrology and Water Behavior

Understanding the environmental factors that influence rescue strategies across various water bodies is crucial for ensuring successful water rescues. Each aquatic environment, be it rivers, lakes, oceans, or floodwaters, presents unique challenges and requires distinct approaches due to its specific characteristics and the dynamic forces at play.

A profound understanding of hydrodynamic principles such as flow rate and water pressure is vital in rescue contexts. The flow rate, typically measured in cubic feet per second (cfs), denotes the volume of water passing through a point in a given time. A swift flow can significantly impede a rescuer's efforts, increasing the complexity and risk involved in reaching those in distress. Conversely, stagnant or slow-moving waters may conceal dangers like soft sediments and sudden drop-offs. Water pressure, on the other hand, affects submerged structures and can determine whether they remain stable under stress. Understanding these elements helps rescuers anticipate potential structural failures and plan for safe extraction routes.

Different water bodies necessitate varied rescue strategies. Rivers, for instance, are characterized by fast currents and often treacherous conditions caused by natural obstacles such as rocks and debris. Rescuers must prioritize swift and accurate assessments of river dynamics to navigate these hurdles effectively. Lakes might appear deceptively calm but pose risks related to thermal gradients and hidden depths. Rescue strategies here should account for thermal protection and the possibility of hypothermia. Oceans introduce additional complexities with tides, waves, and potentially hazardous marine life. In these scenarios, tracking tidal patterns and wave heights becomes essential to timing rescues safely.

Floodwaters present an entirely different set of challenges. Rapidly changing conditions during flash floods can lead to devastating consequences. Recognizing indicators of flash flooding—such as heavy rainfall upstream—can enable preemptive evacuation and preparation, preventing loss of life and property. Moreover, floodwaters may appear shallow but can carry deadly force capable of sweeping away vehicles and people. This inherent unpredictability demands a flexible response framework that incorporates real-time data and continuous monitoring.

Specific hazards like strainers, low-head dams, and undercurrents also significantly influence rescue operations. Strainers, which are objects that allow water to pass but trap larger objects, can be deadly to swimmers caught in their path. Low-head dams create powerful recirculating currents downstream, often described as 'drowning machines' due to their ability to trap victims underneath. Recognizing these hazards allows teams to develop informed tactics to mitigate risks. Undercurrents, though not always visible, exert unseen forces that can rapidly exhaust even strong swimmers. Being cognizant of these factors ensures rescuers approach each mission with caution and respect for the unpredictable nature of water.

The need for specialized training and equipment in these environments cannot be overstated. Rescuers must be adept at reading water conditions, recognizing potential threats, and deploying appropriate rescue techniques. This necessitates rigorous training programs focused on honing skills vital for operating in hostile water settings. Personal protective equipment (PPE), tailored for water rescues, and specialty tools are indispensable for maximizing safety and effectiveness. Implementing advanced communication systems, such as waterproof radios, further facilitates coordination among team members, enabling timely adjustments to strategies as situations evolve.

Risk Assessment and Scene Size-Up

When faced with water rescue scenarios, it's crucial to have a framework for safety evaluation and response tactics. One of the first steps is executing a thorough hazard identification before initiating any rescue efforts. Recognizing the risks associated with swiftwaters, floods, or ice emergencies can significantly impact the outcome of a rescue operation. For instance, in rivers, strainers—obstacles that allow water through but can trap people—pose a significant threat to rescuers and victims alike. By identifying such hazards early, teams can formulate plans that either mitigate these dangers or avoid them entirely.

In conducting hazard assessments, responders should consider the physical environment, weather conditions, and the dynamics of the water. In icy conditions, assessing the stability of the ice is paramount to prevent secondary emergencies during rescue attempts. It's vital for rescuers to rely on their training, experience, and available resources to gauge these threats accurately. Guidelines for identifying hazards before engaging in a rescue can be particularly beneficial. For example, utilizing checklists or mnemonic devices can ensure no potential danger is overlooked.

Once hazards are identified, implementing the 'Go, No-Go' decision-making process becomes essential. This strategy serves as a critical tool in evaluating whether the benefits of a rescue attempt justify the risks involved. The decision hinges on multiple factors, including the conditions of the scene, the responsiveness of the victim, and the readiness of the rescue team. By weighing these elements, responders can make informed choices about whether to proceed with a rescue or adapt their approach.

This decision-making process is not static; it requires continuous reassessment as conditions change. It's crucial to maintain open communication among team members to ensure everyone understands the rationale behind each decision. Often, real-time intelligence gathered from the scene can alter initial assessments, highlighting the need for flexibility and swift judgment calls.

To support effective decision-making and operations management, establishing an Incident Command System (ICS) is fundamental. ICS provides a standardized hierarchy that organizes command, control, and coordination during emergencies. This structure helps streamline operations by clearly defining roles and responsibilities, which reduces confusion and enhances efficiency. An effectively managed ICS facilitates better resource allocation, ensures accountability, and improves communication across all levels of the rescue operation.

For example, assigning specific roles such as incident commander, operations chief, and safety officer can lead to more organized and effective rescues. The incident commander oversees the entire operation, ensuring that strategic objectives are met while prioritizing safety. This level of organization is crucial when dealing with multi-faceted rescue scenarios involving diverse environments like fast-moving floodwaters or unpredictable ice conditions.

Moreover, effective coordination with mutual aid agencies is paramount for comprehensive rescue efforts. Water rescues often require collaboration between various entities such as local fire departments, specialized water rescue teams, law enforcement, and sometimes even the military or Coast Guard. Each agency brings unique skills and resources to the table, enhancing the overall response capability.

Collaboration demands practice and pre-established relationships. Regular joint training exercises can improve interoperability and build trust among different agencies. During actual emergencies, these established connections facilitate smoother communication and cooperation, leading to more efficient rescues. Additionally, having a unified command under ICS can help integrate these diverse teams into a single cohesive force, avoiding jurisdictional conflicts and ensuring all efforts are aligned towards shared goals.

Personal Protective Equipment (PPE) and Rescue Techniques

When it comes to executing safe and successful water rescues, possessing the right gear and understanding effective methods are paramount. These operations often occur in demanding environments where swift currents, floodwaters, and icy conditions pose significant risks. As such, rescuers must be properly equipped and knowledgeable to handle these challenges while ensuring their own safety as well as that of the victims.

Standard Personal Protective Equipment (PPE) is a foundational component of any water rescue operation. This includes Personal Flotation Devices (PFDs) and helmets specifically designed for water rescues. PFDs are crucial, providing buoyancy and reducing the risk of drowning, particularly during turbulent conditions. According to the U.S. Coast Guard's classification, Type III/V PFDs offer adequate buoyancy for swiftwater and ocean rescues, ensuring that responders have the support they need even if swept away by strong currents (Fire Engineering Staff, 1995). Similarly, lightweight helmets with chin straps protect against head injuries from debris or collisions with obstacles.

Beyond basic PPE, specialty equipment becomes indispensable for certain types of rescues. Ice suits, for example, are essential when dealing with frozen water emergencies. These suits provide thermal protection, preventing hypothermia in cold-water environments. Additionally, throw bags are vital tools for river and flood rescues. These bags contain ropes that can be tossed to individuals in distress, allowing them to grab onto the line and be pulled to safety. The usage of throw bags is a testament to the necessity of having versatile gear that can adapt to various rescue scenarios.

In situations where rescuers find themselves at risk, self-rescue tools become critical. For instance, knowing how to position oneself correctly while floating down a river is key. Defensive swimming techniques advise rescuers to float on their backs with feet pointed downstream. This position not only helps avoid hitting obstacles but also provides a clear view of what's ahead, aiding in navigation (DIDIOKMAKINGS, 2025). Furthermore, understanding how to use tensioned diagonal lines—anchored across rivers—facilitates stable crossings in turbulent waters, enhancing both safety and efficiency.

Practical rescue strategies form the backbone of any effective water operation, with the 'Reach, Throw, Row, Go' approach being a cornerstone technique. Each step represents a progression in actions taken during a rescue. Initially, responders should attempt to reach the victim using poles or rescue hooks. If this isn't possible, throwing a lifeline or device such as a throw bag is the next best option. When the situation allows, rowing out to the victim may be necessary, ensuring boats are anchored upstream for stability. Finally, entering the water, only as a last resort, requires proper PPE like dry suits and helmets to maximize safety (DIDIOKMAKINGS, 2025).

To augment the practical use of gear and strategies, it's crucial to evaluate the risk to rescuers versus the benefit of immediate action. Rescuers are trained to assess whether the potential outcome justifies the inherent dangers of attempting a rescue. This decision-making process ensures that safety remains the priority while still striving to accomplish the end goal—saving lives. Assigning roles within the rescue team, such as Incident Commander or Rescue Group Supervisor, further enhances organization and operational efficiency, delineating clear responsibilities and maintaining structured command (Fire Engineering Staff, 1995).

Coordinating with mutual aid agencies is another critical aspect when dealing with large-scale or complex water rescues. Fire departments, EMS, law enforcement, and sometimes the Coast Guard work collaboratively to tackle these emergencies. Seamless communication and cooperation among these entities are essential, facilitating resource sharing and strategic planning. This integrated approach ensures that all available expertise and equipment are leveraged effectively, optimizing rescue outcomes and minimizing risks.

Communication and Coordination Strategies

Effective communication and collaboration are crucial in water rescues, where swift decisions and cohesive teamwork can mean the difference between life and death. In noisy, wet environments, using standardized hand signals and waterproof radios becomes invaluable. These tools provide clear channels of communication, overcoming the obstacles posed by rushing water and disorienting conditions. Hand signals, which remain visible regardless of auditory conditions, offer an unspoken language that rescuers can depend upon. Waterproof radios ensure that verbal communication is possible, allowing teams to relay updates without interruption due to equipment failure. Regular training on these methods ensures proficiency and helps maintain a coherent operation during emergencies.

Another critical aspect of efficient water rescues is integrating multiple agencies and mutual aid responses, especially when large-scale incidents occur. No singular agency can handle every scenario alone; hence, working alongside fire departments, emergency medical services, and volunteer organizations enhances resource availability and expertise. This collaboration demands pre-established relationships and familiarity with each other's protocols, fostering trust and smooth operational flow when incidents arise. Joint training exercises can bridge the gaps between different agencies' procedures, creating a unified front prepared for any situation.

Seamless coordination with external entities like the Coast Guard and law enforcement further amplifies rescue efforts. These entities often have access to specialized equipment, extensive training, and knowledge of local waterways that can prove vital during complex operations. Early involvement and open lines of communication with these authorities allow for the integration of aerial support, maritime navigation, and enforcement capabilities that add layers of safety and efficiency to rescue missions. Maintaining updated contact lists and regularly reviewing coordination plans with them helps establish a reliable framework for collaboration.

Managing logistics in complex operations involving diverse teams and resources involves careful planning and execution. When teams converge on a rescue scene, ensuring that everyone knows their role minimizes confusion and maximizes effectiveness. Establishing a command structure that delineates responsibilities for everything from resource allocation to media management creates a well-organized environment. Equipping teams with personal flotation devices (PFDs), water rescue helmets, and other protective gear such as thermal protection suits for colder waters or ice rescue suits for frigid conditions is non-negotiable. These items not only safeguard rescuers but also boost confidence, allowing them to operate unhindered in treacherous settings.

Furthermore, deploying throw bags, rescue ropes, and carabiners aids in tethered rescues—a common necessity in water emergency scenarios. These tools enable controlled maneuvers and ensure that rescuers can safely reach victims while remaining securely anchored. It is essential that all team members are trained in the proper use and handling of this equipment to prevent mishaps and delays.

Concluding Thoughts

This chapter provided an in-depth exploration of the essential components involved in conducting effective water rescues. By examining the unique challenges presented by swiftwater, flood, and ice environments, it highlighted the necessity for responders to possess a comprehensive understanding of hydrodynamic principles and specialized equipment. Recognizing hazards like strainers and low-head dams, as well as incorporating adaptable rescue strategies, are critical aspects that ensure the safety of both rescuers and victims. The importance of rigorous training was also emphasized, preparing personnel with the necessary skills to navigate these unpredictable situations successfully.

Moreover, the chapter underlined the significance of structured decision-making processes and effective communication within operational teams. Implementing frameworks such as the Incident Command System (ICS) enhances coordination and resource allocation, especially when collaborating with multiple agencies. The integration of personal protective equipment and practical rescue techniques, along with maintaining strong partnerships across various emergency services, fosters a cohesive response effort. These elements collectively contribute to increasing the probability of life-saving outcomes during water-related emergencies, underscoring the vital role of preparedness and teamwork in executing successful rescues.

Chapter 4 Test Questions and Answers

Why are water rescues considered among the most demanding emergency response operations?

Water rescues are complex and dangerous due to unpredictable environments, strong currents, rapidly changing conditions, and the need for specialized training and equipment.

What are the three primary types of water rescues, and how do they differ?

Swiftwater rescues involve fast-moving waters like rivers and rapids. **Flood rescues** occur in widespread, often unpredictable floodwaters that may carry debris. **Ice rescues** involve frozen water surfaces, which can break unexpectedly.

What factors make flood rescues particularly challenging?

Floodwaters can cover large areas, change rapidly, contain dangerous debris, and create strong, unpredictable currents.

What is the importance of historical case studies in water rescues?

Past incidents like Hurricane Florence and Winter Storm Riley provide lessons on preparedness, specialized training, and adaptability.

What is flow rate, and why is it important in water rescues?

Flow rate, measured in cubic feet per second (cfs), determines the speed and volume of water moving, impacting the difficulty and approach of a rescue.

How do different water environments pose unique challenges for rescuers?

Rivers have fast currents and obstacles, lakes can conceal temperature shifts and deep areas, oceans introduce tides and waves, and floodwaters can rapidly intensify.

What is a strainer, and why is it dangerous in water rescues?

A strainer is an obstacle that allows water to pass but traps solid objects, including people, making it extremely hazardous.

What makes low-head dams particularly dangerous?

They create powerful recirculating currents that can trap and drown victims, making them known as "drowning machines."

What is the 'Go, No-Go' decision-making process in water rescues?

It evaluates whether a rescue attempt is feasible based on scene conditions, victim status, and rescuer safety.

Why is continuous risk assessment critical in water rescues?

Water conditions change rapidly, and rescuers must constantly reassess to ensure safety and effectiveness.

What is the role of an Incident Command System (ICS) in water rescues?

ICS provides a structured chain of command, ensuring efficient coordination, resource management, and safety.

Why is coordination with mutual aid agencies essential in water rescues?

Agencies like fire departments, water rescue teams, and law enforcement bring specialized skills and equipment, improving overall response.

Why is specialized training crucial for water rescues?

Rescuers must master technical skills, understand water behavior, and be physically prepared to operate in hazardous conditions.

What are the key personal protective equipment (PPE) items for water rescues?

Personal Flotation Devices (PFDs), water rescue helmets, dry suits, ice suits, and thermal protection gear.

What are throw bags, and how are they used in water rescues?

Throw bags contain a rope that rescuers can toss to a victim, allowing them to be pulled to safety.

What is the 'Reach, Throw, Row, Go' rescue method?

It prioritizes **reaching** for a victim first, then **throwing** a flotation device, **rowing** a boat if needed, and **going** in the water as a last resort.

Why must rescuers monitor weather conditions during a water rescue?

Sudden weather changes can alter water levels, currents, ice stability, and overall scene safety.

How do breached levees and broken storm drains impact urban flood rescues?

They can rapidly increase water levels and current strength, making rescues more dangerous and unpredictable.

Why is thermal protection necessary in ice rescues?

Cold water exposure can lead to hypothermia, reducing a victim's ability to move and increasing the risk of drowning.

How does understanding water pressure aid in rescue operations?

It helps rescuers predict structural failures and determine safe entry and exit points in submerged environments.

Chapter 5: Structural Collapse: Navigating Rubble and Confined Spaces

Navigating the rubble and confined spaces of structural collapses requires a unique blend of knowledge, technique, and intuition. Structural collapse rescue operations are inherently complex and demand that responders approach each situation with an arsenal of strategies tailored to address the specific challenges at hand. The ability to differentiate between various types of collapses—whether partial or total—is crucial for forming effective response plans. Each scenario presents distinct variables and potential hazards that must be considered with precision and foresight.

In this chapter, readers will explore the intricacies of structural collapse rescues, beginning with an overview of different collapse scenarios and their implications for rescue missions. The lessons learned from historical incidents, such as the 9/11 attacks, provide valuable insights into how responders can improve coordination and communication efforts during emergencies. Additionally, the text delves into the importance of adhering to established safety standards like those outlined by the NFPA and highlights the need for continuous training and mental resilience among rescuers. Modern technology's role in enhancing situational awareness and inter-agency collaboration is also examined, alongside the integration of innovative practices that assist in reducing risks and optimizing resource deployment. Through understanding these critical elements, first responders, firefighters, and students in emergency management can prepare themselves better for the demanding task of structural collapse rescue operations, ultimately saving more lives and minimizing risks in high-stress environments.

Understanding Structural Collapse Rescue

Structural collapse rescue is a multifaceted endeavor that requires an in-depth understanding of various collapse scenarios and the strategies to tackle them. One key aspect is differentiating between partial and total collapses, which are crucial for planning effective rescue operations.

Partial collapses involve sections of a structure failing while other parts remain intact. They often occur due to localized factors such as structural weaknesses or minor seismic activity. The implications of a partial collapse include the potential for secondary collapses, making it critical for rescuers to assess stability continuously. In contrast, total collapses are more catastrophic, involving the entire structure's failure. This could result from significant events like earthquakes or large-scale explosions. Understanding the nature of these collapses helps responders anticipate the extent of debris and victim locations within the rubble.

Analyzing historical incidents provides valuable insights into structural collapse dynamics. The 9/11 terrorist attacks serve as a profound case study, highlighting both challenges and lessons learned. The complexity of the World Trade Center collapse required unprecedented coordination among agencies. Rescue attempts were complicated by hazardous debris, fires, and unstable structures. This tragic event emphasized the need for comprehensive incident command systems and enhanced inter-agency communication during large-scale emergencies.

The challenges faced during structural collapse rescues are numerous, with time sensitivity being paramount. Rescuers work against the clock to locate and extricate victims, as survival rates decrease significantly after the first 24 hours. Rapid assessment of the scene is essential to prioritize search areas and deploy resources efficiently. Safety risks are prevalent, ranging from falling debris to hazardous materials exposure. Rescuers must be equipped with proper protective gear and follow strict protocols to minimize these dangers.

Adhering to established standards, such as the NFPA 1006 and 1670, is vital to ensuring effective Urban Search and Rescue (USAR) operations. NFPA 1006 outlines professional qualifications for rescuers, emphasizing skills in areas like rope rescues and confined space operations. Meanwhile, NFPA 1670 provides guidelines for organizational management, covering everything from training requirements to equipment specifications. Following these standards ensures that responders are well-prepared and capable of executing complex rescues safely.

In addition to technical skills, mental resilience is crucial for rescuers dealing with high-stress environments. Structural collapses can pose psychological challenges due to traumatic scenes and the pressure to save lives. Organizations should prioritize mental health support and provide training that includes stress management techniques.

Overall, defining and exploring the dynamics of structural collapse rescue involves understanding different collapse types, learning from past incidents, recognizing inherent challenges, and adhering to safety standards. By mastering these elements, rescue teams can improve their effectiveness and increase the chances of saving lives during these critical operations. Thus, continuous training and adherence to regulatory guidelines become pivotal elements in honing the skills necessary for successful intervention in structural collapse scenarios (<i>Federal Register :: Request Access</i>, n.d.).

Building on the framework provided by NFPA standards, rescuers must also employ modern technology and innovative practices. For instance, drones equipped with cameras and sensors offer real-time aerial views of collapse sites, enhancing situational awareness without exposing personnel to unsafe conditions. These technologies facilitate efficient resource allocation and risk mitigation.

Furthermore, fostering inter-agency collaboration is integral to successful rescue operations. Establishing strong partnerships among local, state, and federal emergency services ensures a unified response when a structural collapse occurs. Joint training exercises and shared communication platforms allow these diverse entities to coordinate efforts seamlessly.

To prepare for the unpredictable nature of structural collapses, agencies should invest in research and development. Experimenting with new materials and construction techniques can lead to more resilient buildings, reducing the likelihood of total collapses. Additionally, enhancing public education about building codes and emergency preparedness encourages community resilience.

Causes and Types of Structural Collapse

Understanding the various causes and types of structural collapse is essential for first responders, firefighters, and students in emergency management. Analyzing these factors not only prepares them for rescue operations but also aids in developing effective strategies. This subpoint delves into both natural and human-induced causes, as well as the classification of different collapse types, each presenting distinct challenges during rescue missions.

Natural disasters remain one of the predominant causes of structural collapses globally. Earthquakes, hurricanes, and tornadoes possess the destructive capability to bring down buildings and infrastructure, often leaving communities devastated and overwhelmed. For instance, earthquakes generate powerful ground movements that compromise the integrity of structures, particularly those not designed to withstand seismic forces. The 2010 Haiti earthquake serves as a prime example where widespread building failures led to massive casualties (Terry, 2017). Hurricanes, on the other hand, bring intense winds and flooding that weaken structures, displacing populations and requiring large-scale rescue operations.

Understanding these natural triggers helps in anticipating potential collapses and preparing adequately.

Human-induced failures represent another significant cause of structural collapse, often rooted in poor construction practices or deliberate acts of terrorism. Buildings constructed with inferior materials or without adherence to safety regulations are prone to failure, especially under load stress or adverse weather conditions. Historical incidents like the collapse of the Sampoong Department Store in South Korea highlight the catastrophic impact of negligent construction practices. Furthermore, acts of terrorism, such as bombings, intentionally compromise structural integrity, endangering lives and complicating rescue efforts (Daley, 2018). Recognizing these vulnerabilities underscores the importance of stringent building codes and enhanced security measures.

In terms of classification, structural collapses can manifest in several ways, each type affecting rescue operations differently. The pancake collapse occurs when floors fall directly onto one another due to the failure of load-bearing walls. This type creates compacted layers of debris, posing significant challenges in locating and accessing trapped victims. Rescue teams must exercise caution and deploy specialized equipment to navigate these tight spaces safely (Daley, 2018).

The lean-to collapse involves one side of the structure failing while the opposite remains intact, resulting in a slanted floor position. It provides a void space near the unaffected wall, potentially offering temporary refuge for survivors. Rescuers navigating a lean-to collapse must prioritize stabilizing the remaining structure to prevent further collapse while carefully extracting victims from voids.

A V-shaped collapse typically occurs when heavy loads in the center of a structure cause lower walls or floor joists to fail. This results in two distinct voids along exterior walls, which may harbor survivors. The irregular shape of the debris requires rescuers to adapt their approach, ensuring they account for hidden hazards and maintain communication with trapped individuals.

Cantilever collapses are particularly dangerous, occurring when one or more walls fail while the other ends remain attached to bearing walls. Voids within the debris are sporadic, making it challenging to identify safe pathways for rescue. Adequate shoring is crucial before operations commence to stabilize the precarious conditions and allow safe access to victims. Rescuers must meticulously evaluate the collapse scene to determine the best entry points and methods for victim recovery (Terry, 2017).

Finally, the domino collapse involves sequential failures of structural components, similar to falling dominoes. Such incidents often result in extensive debris fields and unpredictable hazard zones. Rescue efforts require strategic planning and coordination to mitigate risks associated with secondary collapses. Establishing clear communication channels and utilizing updated mapping techniques aid in navigating these complex environments effectively.

Evaluating how each type of collapse presents unique challenges during rescues highlights the importance of targeted training and preparedness. First responders must be equipped with knowledge of different collapse scenarios and possess the necessary skills to adapt rapidly. Collapses demand quick decision-making, innovative problem-solving, and collaboration among rescue teams to maximize survival chances and minimize risks.

Risk Assessment and Incident Management

Conducting initial scene evaluations is a pivotal first step in assessing risk and managing incidents at collapse sites. This process involves a thorough examination of the site to identify potential hazards, which could range from structural instability to hazardous materials present on-site. One essential guideline for this evaluation is to conduct a rapid reconnaissance to assess the structural integrity of buildings, marking them with standardized systems to communicate their status effectively (<i>Collapse Incident Marking Systems for Structural Triage and Rapid Search Reconnaissance Operations</i>, 2002). By understanding the condition of the structures, responders can make informed decisions about the safety of operations and necessary resource allocation.

Next, establishing incident command and coordinating with agencies like FEMA are crucial components of effective incident management. Once the initial evaluation is complete, setting up a unified command structure allows for seamless coordination among various response teams. This framework helps in allocating responsibilities and ensuring that all team members work towards common goals. A cohesive command setup facilitates communication between local responders and federal agencies such as FEMA, which may provide additional support and resources during large-scale emergencies (Lee, 2007).

Implementing safety measures against secondary collapses and environmental hazards is another critical aspect of managing incidents at collapse sites. Secondary collapses pose significant risks to both victims trapped within structures and the rescue teams working to free them. As part of risk mitigation, it is vital to secure utilities such as gas and electricity swiftly to prevent fires or explosions, an action emphasized in early hazard assessments (Lee, 2007). Additionally, establishing perimeters and rerouting civilian traffic help in minimizing disturbances that could lead to further instability.

Utilizing a structural triage system is instrumental in prioritizing search areas effectively. This system involves a three-step process: identifying affected buildings, conducting rapid assessments, and determining the necessary level of detailed evaluation required for specific structures. By prioritizing which areas require immediate attention, rescue teams can focus their efforts on locations with the greatest likelihood of finding survivors. The triage system also aids in ensuring that resources are allocated efficiently, preventing duplication of efforts and maximizing the chances of successful rescues (<i>Collapse Incident Marking Systems for Structural Triage and Rapid Search Reconnaissance Operations</i>, 2002).

Search and Rescue Techniques

Effective search and rescue operations in structural collapse scenarios require a systematic approach to navigate the complex environment of rubble and confined spaces. Understanding the three primary phases of such operations—surface search, void search, and technical search—is crucial for first responders.

The surface search phase is the initial step, focusing on quickly identifying visible victims or signs of life without delving deep into unstable structures. Speed and organization are essential to maximize the chances of early rescues (Marcus, 2024). This phase often involves visual scans and basic tools to assess easily accessible areas. Deploying specialized K9 teams enhances this process significantly. Trained dogs are vital assets as they can swiftly detect human scents even in chaotic and dust-filled environments, which may be challenging for electronic devices (Hu et al., 2019).

Once surface assessments indicate possible survivors, the void search phase begins. This phase targets spaces within collapsed structures where survivors might be trapped but not immediately visible. The use of advanced listening and detection tools is critical here. Seismic sensors detect the faintest vibrations caused by movements or sounds made by conscious survivors (Marcus, 2024). These devices provide rescuers with accurate information about potential victim locations, allowing for more focused efforts and minimizing unnecessary disturbances to hazardous structures.

Technical search represents the most advanced phase of collapse response, involving a thorough and strategic exploration of the debris. It demands the deployment of sophisticated equipment to locate deeply buried victims. Such technology includes ultra-wideband radar and USAR cameras, which penetrate thick layers of rubble and provide real-time imagery of victim locations (Marcus, 2024). These tools complement K9 units by providing visual confirmation of life signals initially detected by scent. Thermal imaging is another vital tool during technical search, effectively identifying heat signatures even in low-visibility conditions like smoke or darkness. Its capacity to pinpoint body heat helps direct rescuers precisely to where survivors might be.

Quick and accurate search methods are paramount in determining the success of these operations. In the chaotic aftermath of structural collapses, time is an adversary, with survival rates rapidly decreasing as hours pass (Hu et al., 2019). Thus, the prompt initiation of all search phases, particularly by experienced teams, is necessary to save lives. Ensuring that each phase builds on the previous one enhances the efficiency of the entire operation, making certain that no potential area remains unexplored.

Moreover, employing the right equipment and technologies is indispensable. Tools must meet standards for reliability and durability in extreme conditions, ensuring continuous functionality regardless of environmental challenges. Equipping rescue teams with these tools while maintaining constant training on their usage ensures readiness for immediate deployment when disaster strikes.

Extrication, Stabilization, and Safety Measures

In the high-stakes arena of structural collapse rescue operations, mastering techniques for victim extrication and ensuring site stability and safety is paramount. As rescuers navigate these perilous situations, their skills in handling breaching, breaking, and shoring systems are crucial for successful outcomes. This section will explore various methods and guidelines crucial to these operations.

When confronted with collapsed structures, applying appropriate breaching and breaking methods without triggering secondary collapses is essential. Rescuers must analyze the structure's current state and choose the safest path to reach trapped victims. The goal is to create access openings by strategically breaching through walls, ceilings, or floors, while maintaining structural integrity (<i>Building Collapse Rescue: Operational Considerations</i>, 2019). Traditional tools such as sledgehammers, bolt cutters, and saws may be employed, alongside modern equipment like pneumatic or hydraulic tools that allow precision cutting while minimizing vibration and the risk of further destabilization (<i>Structural Collapse Flashcards</i>, 2025).

Utilizing equipment designed for lifting heavy debris and stabilizing structures plays a pivotal role in managing site safety. Machinery such as cranes and backhoes can efficiently remove larger sections of debris, provided expert rigging practices are in place to control movement and loading stresses on adjacent structures (<i>Building Collapse Rescue: Operational Considerations</i>, 2019). Additionally, employing smaller, maneuverable tools that facilitate working within confined spaces—such as air bags or jacks—enables teams to gradually elevate heavy slabs, preventing sudden shifts that could endanger victims or responders.

In these operations, victim stabilization is an immediate priority to prevent injury exacerbation during removal. Constrained spaces demand careful packaging of victims using stretchers or boards tailored for confined environments. Rescuers must ensure the patient's spine and limbs are immobilized when necessary, employing straps and padding to maintain alignment and reduce discomfort. Packing materials, such as blanket rolls or small wedges, can further secure victims and enable smooth extrication through narrow apertures (<i>Structural Collapse Flashcards</i>, 2025).

Implementing effective shoring systems and temporary supports is another critical aspect. Shoring serves as a preventative measure against further collapses, providing temporary stability to structurally weakened areas. Fire service agencies utilize various systems including wood, mechanical, pneumatic, and hydraulic options depending on site-specific requirements. Wood shores, kept short and robust, are frequently combined with pneumatic or hydraulic alternatives to enhance support capacity (<i>Building Collapse Rescue: Operational Considerations</i>, 2019). A systematic approach to evaluating each void space from multiple angles (top, bottom, sides) before shoring ensures comprehensive stability.

Adopting operational guidelines is fundamental to managing the complex dynamics at collapse sites effectively. Teams must designate a safety officer responsible for assessing ongoing risks and swiftly intervening if stability is compromised. Establishing clear communication channels and incident command zones—distinct 'hot', 'warm', and 'cold' areas—facilitates coordination and resource allocation (<i>Building Collapse Rescue: Operational Considerations</i>, 2019). These predefined zones not only delineate areas of varying hazard levels but also dictate where specific activities, such as equipment staging and personnel assembly, should occur.

Recognizing the unpredictable nature of structural collapse events, preparation should extend beyond immediate actions. Fire departments must maintain updated resource lists incorporating specialized personnel and equipment from external providers when agency resources are insufficient. This proactive measure reduces delays and enhances operational efficiency by guaranteeing timely availability of essential assets, from technical advisors to heavy machinery (<i>Building Collapse Rescue: Operational Considerations</i>, 2019).

A successful rescue operation hinges on a detailed understanding of building layouts and potential access routes. Reconnaissance teams perform six-sided surveys, assessing the top, bottom, and all sides of affected structures. Through visual inspection and auditory cues, these teams locate possible entry points and prioritize efforts based on victim visibility and access feasibility (<i>Structural Collapse Flashcards</i>, 2025).

Bringing It All Together

In this chapter, we explored the intricate nature of structural collapse rescue operations. The focus was on understanding different types of collapses, such as partial and total, which require detailed assessment to plan effective rescues. We examined both natural disasters and human-induced causes that lead to these scenarios, emphasizing the importance of preparedness and adherence to safety standards. Through historical analyses, such as the 9/11 attacks, we gained insights into improved coordination and communication among agencies. The chapter also highlighted the challenges of time-sensitive operations, stressing the need for rapid assessments, careful prioritization, and resource allocation. By following established guidelines like NFPA standards, and maintaining mental resilience, rescuers are better equipped to handle high-stress environments effectively.

Continuing from this foundation, advanced technology and inter-agency collaboration emerged as essential components in enhancing response strategies. Tools like drones and seismic sensors improve situational awareness, while joint training exercises foster seamless cooperation between various emergency services. The chapter underscored the value of research and development in building more resilient structures, aiming to reduce future collapse risks. Moreover, educating communities about building codes and emergency preparedness contributes to overall resilience. Ultimately, a comprehensive grasp of these elements strengthens the capability of first responders, firefighters, and students in emergency management, equipping them with the knowledge needed to navigate complex rescue operations successfully.

Chapter 5 Test Questions and Answers

What is the key difference between partial and total collapses?

A partial collapse involves sections of a structure failing while other parts remain intact, whereas a total collapse results in the complete failure of the structure.

Why is understanding the nature of collapse scenarios important for rescuers?

It helps responders anticipate debris distribution, victim locations, and potential hazards, allowing for more effective rescue strategies.

What are some primary causes of structural collapses?

Natural disasters (earthquakes, hurricanes, tornadoes), poor construction practices, and deliberate acts such as terrorism.

How does a pancake collapse differ from a lean-to collapse?

A pancake collapse occurs when floors stack on top of one another due to load-bearing wall failure, while a lean-to collapse happens when one side of the structure remains intact, creating a void space.

What are the dangers associated with a V-shaped collapse?

It creates two distinct voids along exterior walls, which may contain survivors but also pose risks due to unstable debris.

What made the 9/11 structural collapse particularly challenging for rescuers?

The complexity of the debris, hazardous materials, fires, and the need for inter-agency coordination made rescue operations extremely difficult.

What lessons were learned from the Haiti earthquake in terms of structural collapse response?

The importance of earthquake-resistant buildings, rapid deployment of international aid, and organized search and rescue efforts.

How did the collapse of the Sampoong Department Store in South Korea highlight the risks of poor construction practices?

The failure was attributed to structural overload and design flaws, emphasizing the need for strict building codes and inspections.

Why is time sensitivity crucial in structural collapse rescues?

The survival rate of trapped victims decreases significantly after the first 24 hours.

What role does an incident command system play in managing collapse rescue operations?

It ensures coordinated efforts among local, state, and federal agencies, streamlining communication and resource allocation.

How does a structural triage system aid in rescue operations?

It prioritizes search areas based on the likelihood of survivor presence and the structural integrity of buildings.

What are secondary collapses, and why are they a major concern?

Secondary collapses occur when unstable debris shifts during rescue efforts, endangering both victims and rescuers.

What are the three primary phases of search and rescue operations in structural collapses?

Surface search, void search, and technical search.

How do K9 search units enhance rescue operations?

Trained dogs can detect human scents under rubble, increasing the chances of locating survivors quickly.

What technology is used in technical search operations?

Seismic sensors, ultra-wideband radar, USAR cameras, and thermal imaging to detect survivors within debris.

Why is shoring important in structural collapse rescue?

Shoring stabilizes weakened structures, preventing further collapses while rescuers work to extract victims.

What are breaching and breaking techniques used for?

These techniques create safe access points through debris without causing additional structural failures.

What personal protective equipment (PPE) is essential for rescuers?

Helmets, gloves, respiratory protection, eye protection, and reinforced boots to protect against falling debris and hazardous materials.

What NFPA standards guide Urban Search and Rescue (USAR) operations?

NFPA 1006 (professional qualifications for rescuers) and NFPA 1670 (organizational management and training).

How can agencies improve resilience against structural collapses?

By investing in advanced training, modern technology, inter-agency collaboration, and research into stronger building materials.

Chapter 6: Vehicle Extrication: Strategies for Saving Lives in Auto Accidents

Vehicle extrication is a vital procedure in auto accident response, often determining the difference between life and death for those involved. The process demands not only technical skill but also a deep understanding of diverse and challenging scenarios that first responders face daily. With vehicles constantly evolving through advancements in technology and design, the strategies employed in extricating individuals must adapt to meet new challenges head-on. First responders are tasked with executing these operations swiftly and safely, balancing the urgency of the situation with the meticulous precision required to avoid additional harm.

This chapter delves into the core strategies and techniques integral to vehicle extrication. Readers will explore the importance of rapid assessment and decision-making by first responders at the scene of an accident, emphasizing the role each second plays during the pivotal 'Golden Hour.' The text covers various obstacle management tactics, particularly focusing on modern vehicle designs that complicate traditional rescue efforts. It further outlines the necessity for specialized equipment and up-to-date training tailored to contemporary vehicle features, ensuring responders are capable of overcoming any barriers they encounter. Additionally, the collaboration between different emergency services and continuous education and mental health support for responders are highlighted as fundamental to successful operations. Through these insights, the chapter aims to equip readers with a comprehensive understanding of how to effectively navigate the complexities of vehicle extrication, ensuring maximum survivability in critical situations.

Introduction to Vehicle Extrication

Understanding the foundational elements of vehicle extrication is crucial as these actions often dictate life or death outcomes, especially within the critical time frame known as the 'Golden Hour.' This term refers to the first 60 minutes after a traumatic injury, during which prompt medical treatment significantly improves survival chances. First responders, such as firefighters and emergency medical services, are typically the initial on-scene professionals tasked with rescuing individuals trapped in vehicles following accidents. The decisions made during these early moments are pivotal and necessitate a comprehensive understanding of extrication techniques.

First responders are trained to assess situations quickly, determining the best course of action amid often chaotic accident scenes. They must navigate a myriad of challenges including scene hazards like traffic, fire, and fuel leaks. These dangers require swift and effective decision-making to ensure both victim and responder safety. Traffic is a common hazard, as ongoing flow around an accident site can create dangerous conditions for responders working to extricate victims. Fire hazards arise from fuel leaks and damaged electrical systems, potentially leading to explosions or rapidly spreading fires. Addressing these hazards requires establishing secure perimeters and controlling the environment before commencing extrication.

Moreover, modern vehicle designs, characterized by advanced technology, present additional obstacles that have transformed traditional rescue operations. Vehicles now contain components such as high-strength metals, extensive electronic systems, and alternative power sources like electric batteries and hybrid engines. Such advancements make it imperative for first responders to stay updated with current vehicle construction techniques and incorporate new strategies for effective extrication.

High-strength metal in car frames and structures often poses significant challenges to standard extrication tools. Rescue teams must be equipped with specialized equipment capable of cutting through these reinforced materials. For instance, ultra-high-strength steel (UHSS), commonly employed in newer vehicle models, may resist conventional cutting tools, prompting the need for hydraulic cutters designed specifically for this purpose (Loone, 2020).

Additionally, the increased prevalence of electric and hybrid vehicles introduces risks such as electrocution and specific considerations related to battery handling. Identifying and disabling electrical systems safely is crucial to avoid potential hazards during rescue operations. Training programs for first responders, therefore, increasingly emphasize awareness and management of these vehicular characteristics. It's vital for training curricula to integrate classroom learning with hands-on practical experience to fully prepare responders for the complexities of modern rescue scenarios (Bossert, 2024).

The interaction between first responders and emergency medical services (EMS) personnel is another cornerstone of successful vehicle extrication. While firefighters focus on accessing and removing victims, EMS providers deliver immediate medical care, stabilizing patients before and after extraction. Collaboration ensures that the patient receives comprehensive care throughout the entire process, minimizing the risk of further injury during extrication (Bossert, 2024).

To counter the evolving challenges associated with vehicle extrication, continuous education and interdisciplinary coordination are necessary. Advancements in automotive technology mean that extrication tactics must be regularly updated and practiced to reflect the latest industry developments. Joint training exercises involving different emergency response teams foster improved communication and enhance operational efficiency. By staying informed and adaptive, first responders can more effectively manage the risks posed by the changing landscape of automobile accidents.

The commitment to ongoing education also extends to understanding the psychological impact of rescue operations on both victims and responders. High-intensity situations coupled with exposure to traumatic scenes can affect mental health. Providing support mechanisms such as counseling and stress management resources can mitigate these effects, promoting resilience among rescue teams.

Evaluating Different Types of Vehicle Accidents

In the critical and urgent world of vehicle extrication, understanding how different types of vehicle accidents influence rescue strategies plays an essential role in determining the outcomes for those involved. Each type of collision presents unique challenges that necessitate specific approaches to extrication efforts, impacting the efficiency and effectiveness of first responders on the scene.

Frontal collisions are one of the most common and visually impactful types of vehicle accidents. They often result in dashboard intrusions and steering wheel compression injuries. The force generated during such impacts can drive parts of the vehicle's interior into passengers, causing severe trauma. For first responders, this scenario requires precise extrication techniques. Tools like dash rolls and hydraulic spreaders are employed to carefully push away or cut through metal panels, thereby freeing trapped victims without exacerbating their injuries. Understanding the mechanics of frontal collisions enables rescuers to predict likely points of entrapment and plan efficient routes for victim access and extraction.

Side-impact crashes, also known as T-bone collisions, present another complex challenge for emergency personnel. These incidents typically lead to significant structural deformation of the vehicle, often resulting in severe occupant entrapment. The impact primarily affects the doors and side panels, compressing vulnerable areas and restricting movement. The nature of these collisions necessitates specialized equipment, such as side airbags, cutters, and spreaders, to peel away layers of twisted metal. Statistically, occupants seated on the impacted side face increased risks due to the proximity and direct exposure to the crash site (Otte, 1984). Furthermore, narrow fixed object impacts prove more hazardous than vehicle-to-vehicle crashes, adding to the complexity (Malliaris, 1992). By analyzing data from safety studies and databases such as NASS and WLIRC, first responders can gain insights into injury patterns and refine their methods accordingly, ensuring the best possible care for the injured parties (Augenstein et al., 2025).

High-speed crashes bring an elevated level of risk both in terms of injury severity and vehicle destruction. At greater speeds, the force of impact generates extensive damage to the vehicle, complicating rescuer access and increasing the likelihood of debris hazards. Such accidents demand immediate action, prioritizing not just the physical extrication but also the stabilization of the victims due to heightened medical emergencies. Protocols during high-speed extrications must be meticulously planned, with continuous communication among team members to manage the chaotic scene effectively. Understanding the dynamics of high-speed impacts allows rescuers to anticipate potential complications and prepare adequately, ensuring the timely delivery of life-saving interventions.

Accidents involving large vehicles, such as trucks or buses, introduce additional variables into the rescue operation. Larger vehicles have increased weight and may carry hazardous cargo, posing additional risks to rescuers. These situations call for strategic planning and sometimes require high-angle rescue techniques due to the height and structure of the involved vehicles. Extricating individuals from such massive structures demands coordinated use of heavy-duty equipment like cranes or jacks capable of lifting thousands of pounds safely. Additionally, those responding to such incidents must account for potential chemical or fuel leaks and be equipped to handle such exigencies swiftly. Recognizing the complexities associated with large vehicle accidents helps responders implement appropriate safety measures and maximize efficiency.

The severity and variation of vehicle accidents underscore the need for ongoing training and adaptation within the field of vehicle extrication. Familiarity with cutting-edge equipment and techniques is paramount, enabling first responders to navigate diverse scenarios effectively. Moreover, sharing knowledge and insights from past incidents fosters a culture of continuous improvement, driving advancements in rescue operations and enhancing overall outcomes.

Conducting Effective Scene Size-Up and Risk Assessment

Mastering scene evaluation techniques is a critical skill for first responders charged with conducting vehicle extrication during auto accidents. The foremost goal in extrication operations is to ensure the safety and efficiency of the procedures, which starts with properly evaluating the accident scene. This subpoint delves into several key components that first responders should consider when approaching a vehicle accident scene.

One of the primary steps in scene evaluation is establishing a safe zone, which involves controlling the surrounding traffic. Traffic control ensures that emergency personnel can work without the threat of being struck by passing vehicles. It also protects the victims and other civilians who might be present at the scene. To implement effective traffic control, responders may use cones, flares, or other signaling devices to direct traffic away from the site of the incident. Additionally, coordinating with law enforcement to enforce road closures or diversions can further secure the area (<i>Chapter 38: Vehicle Extrication and Special Rescue Flashcards</i>, 2025).

Once a safe perimeter has been established, a thorough 360-degree assessment of the crash site is essential. This type of assessment allows responders to better understand the scope of the incident by identifying the number of victims involved, the extent of their entrapment, and the stability of each vehicle. By walking around the scene and inspecting vehicles from all angles, responders can gather crucial information about how to approach the extrication process. For instance, recognizing signs of instability in a vehicle's positioning, like leaning or potential rolling hazards, dictates the stabilizing measures needed before any further action can occur (<i>Chapter 38- Vehicle Extrication and Special Rescue Flashcards</i>, 2025).

Coordinating with an incident command center is another important aspect of effective scene evaluation. Incident command serves as the hub of communication and decision-making, especially in complex multi-agency responses. Firefighters, police officers, paramedics, and other specialized rescue teams need to clearly communicate and collaborate to avoid duplication of efforts and miscommunication. A designated incident commander typically assigns roles, oversees logistics, and provides updates throughout the operation. Successfully managed incident command ensures that resources are used efficiently and that everyone involved in the response is informed and aligned with the rescue strategy.

As part of the scene evaluation, first responders must also identify any secondary hazards that could pose additional risks. These hazards include downed power lines, fuel leaks, or hazardous materials that could threaten the scene's safety. For example, if power lines are found near a crashed vehicle, responders should instruct occupants to remain in the vehicle until the lines have been secured by utility personnel. Similarly, identifying and addressing fuel leaks is vital to prevent ignition and subsequent fires (<i>Chapter 38- Vehicle Extrication and Special Rescue Flashcards</i>, 2025).

Evaluating these potential hazards helps responders mitigate threats that could complicate the extrication process or endanger those involved. Quick and accurate identification followed by appropriate action — such as roping off contaminated areas or ensuring utilities are cut off — can significantly reduce the risk of secondary incidents. Overall, mastering scene evaluation techniques requires knowledge, practice, and vigilance. First responders play an invaluable role in saving lives by systematically assessing and managing the chaotic environments they encounter during vehicle extrications.

Employing Vehicle Stabilization Techniques

In the intricate process of vehicle extrication, stabilization plays a pivotal role in ensuring both victim safety and responder security. Implementing effective stabilization techniques is crucial to prevent secondary movements that could compromise the rescue operation and exacerbate injuries. To begin with, understanding the foundational aspects of stabilization—such as chocking wheels and securing suspension systems—is essential for preventing vehicle displacement during the extrication process.

Chocking the wheels is perhaps one of the simplest yet most effective methods in stabilizing a vehicle. By placing wheel chocks behind or in front of the tires, rescuers can effectively prevent the vehicle from rolling forward or backward, thus minimizing any risk of additional injury or accident at the scene. Moreover, securing the suspension system can add another layer of stability, particularly in vehicles that might be precariously balanced or positioned on uneven terrain. This involves using tools like tie-down straps or ratchet straps to further immobilize the vehicle's movement.

Beyond these basic techniques, specialized tools such as cribbing, struts, and hydraulic stabilizers provide the necessary support for more complex or unstable scenarios. Cribbing involves stacking blocks, typically made of wood or plastic, to create a stable base under the vehicle. This method not only aids in supporting the weight of the vehicle but also allows responders to use other extrication tools safely. Struts, on the other hand, are versatile stabilizing devices that can be adjusted to fit different angles and positions. Junkyard Dog Industries, for instance, offers rescue struts that are lightweight and easy to deploy, facilitating quicker stabilization (<i>Extrication Cribbing & Stabilization</i>, 2018). These struts accommodate conventional and ultra-modern vehicle construction features, making them indispensable in modern rescue operations.

Hydraulic stabilizers are another vital component in the stabilization toolkit. They offer the advantage of being able to exert significant pressure, enabling rescuers to lift or pull parts of the vehicle without relying solely on manual strength. In highly unstable situations, such as when the vehicle is resting on its side or roof, these tools become even more critical. For side-resting and roof-resting vehicles, special considerations must be accounted for to ensure stability. A buttress system or modified raker shore with multiple contact points often works well for these scenarios (<i>TR vr STABILIZATION EXTRICATION Flashcards</i>, 2025). Additionally, voids should be shimmed to prevent any unintentional shifts, while ground base plates should be anchored securely.

Water rescue scenarios introduce a unique set of challenges due to fluid dynamics. Traditional stabilization methods may prove less effective as water can significantly alter the vehicle's position unpredictably. In such cases, buoyancy becomes a concern, and extra care must be taken to anchor the vehicle underwater. Specialized floatation and anchoring systems may need to be employed to maintain control over the vehicle until victims can be safely extricated. It's crucial that responders adapt their techniques to account for these dynamic conditions, considering factors such as the depth of water and current speed.

The selection and application of stabilization tools must always consider the specific situation at hand. This means assessing the degree of instability, identifying potential hazards, and selecting the appropriate equipment to mitigate risks effectively. Safety remains paramount, as an unstable vehicle poses threats not just to trapped individuals but also to rescue personnel operating around it.

Moreover, rescuers must be trained to recognize the various types of vehicle movements—vertical, horizontal, pitch, roll, and yaw—and understand how each can impact stabilization efforts. For example, vertical movement can cause a vehicle to sink into softer ground, while horizontal shifts might indicate an underlying imbalance. By predicting these movements, responders can preemptively employ countermeasures to secure the site safer.

Advanced Extrication Tools and Tactical Approaches

Vehicle extrication is a critical skill for first responders, requiring the adept use of specialized tools and techniques. Among the fundamental methods are the hydraulic tools such as the Jaws of Life, which have transformed rescue operations since their introduction in the 1960s (<i>Rescue Tools: Pros and Cons of 3 Power Sources</i>, n.d.). These tools offer substantial cutting and spreading power essential for quickly accessing victims trapped in vehicles. Despite their strengths, hydraulic systems have certain drawbacks, primarily their weight and reliance on hydraulic fluid. Their effectiveness diminishes under extreme temperatures, necessitating careful maintenance and preparation before deployment.

Pneumatic tools complement hydraulic systems by providing portability and ease of operation (<i>Rescue Tools: Pros and Cons of 3 Power Sources</i>, n.d.). Airbags, a key pneumatic tool, are indispensable for lifting heavy vehicles or stabilizing them during extrication. They enable responders to gently elevate vehicles, ensuring minimal further injury to trapped occupants. However, these tools depend on a reliable air supply and may struggle with thicker vehicle structures. It's crucial that first responders understand the limitations and optimal applications of pneumatic tools to employ them effectively on scene.

The rise of electric and hybrid vehicles adds complexity to extrication scenarios. These vehicles require special attention due to their high-voltage systems, which can be hazardous if not properly managed. Responders must be trained in shutting down these systems safely to avoid electrical hazards during rescues. This involves identifying and disconnecting the battery and ensuring that all circuits are inert before commencing extraction efforts. Understanding the layout and unique characteristics of electric vehicles is vital for safe and efficient operations, as conventional approaches might not always apply.

When it comes to specific extrication techniques, methods like dash roll and B-post blowouts are essential for gaining access to victims in complex entrapments. The dash roll involves displacing the dashboard away from the patient, typically using hydraulic rams, thereby freeing trapped legs or bodies. This technique demands precision and understanding of the vehicle's structure to execute successfully without causing additional harm. Likewise, B-post blowouts allow responders to remove vehicle doors swiftly, clearing paths for victim extrication. Mastery over these techniques not only improves efficiency but also enhances the chances of survival for those involved in severe collisions.

Maintaining a diverse array of tools and techniques ensures readiness for various extrication challenges. Hydraulic tools remain the backbone for their unparalleled power, especially in dealing with robust vehicle frames. Meanwhile, pneumatic options offer versatility and quieter operation, making them ideal for specific tasks where traditional hydraulic solutions might be cumbersome or unnecessary. As technology marches forward, incorporating newer electric tools into the arsenal provides yet another layer of adaptability, though it requires thorough training and understanding of the power limitations inherent in battery-operated devices.

Training forms the bedrock of successful extrication operations. Firefighters and first responders must undergo rigorous practice to become proficient in both tool usage and tactical approaches. Training sessions should emphasize the nuances of operating under pressure, recognizing when and how to deploy each type of tool based on the scenario at hand. Regular drills replicating real-world conditions help ensure that rescuers are prepared to act decisively and effectively when confronted with life-and-death situations.

Equally important is the maintenance of all equipment used in extrication. Routine checks and servicing prevent mechanical failures that could impede rescue efforts. Keeping tools in peak condition guarantees their reliability during critical moments, allowing responders to focus solely on the task without distractions about equipment performance. Proper storage and handling protocols further enhance tool longevity and operational efficacy.

Lessons Learned

In this chapter, we delved into the critical intricacies of vehicle extrication, focusing on strategies and techniques essential for maximizing survivability during auto accidents. We explored the challenges first responders face, from traditional hazards like traffic and fire to contemporary complications arising from modern vehicle designs equipped with high-strength materials and advanced technology. The emphasis on continuous education and interdisciplinary coordination was highlighted as a key factor in evolving rescue strategies, underscoring the importance of staying updated with automotive advancements to ensure effective extrication operations.

Furthermore, we examined different types of vehicle accidents, such as frontal collisions, side-impact crashes, and high-speed incidents, each necessitating unique approaches for successful rescues.

The significance of scene evaluation, stabilization techniques, and the use of advanced extrication tools were also discussed, reinforcing the meticulous preparation required for dealing with varying scenarios. By equipping first responders with comprehensive knowledge and skills, this chapter aims to enhance their capability to manage complex auto accident situations effectively, ensuring victim safety while maintaining responder security.

Chapter 6 Test Questions and Answers

What is the "Golden Hour" in vehicle extrication?

The "Golden Hour" refers to the first 60 minutes after a traumatic injury, during which prompt medical treatment significantly improves survival chances.

Who are the first responders typically involved in vehicle extrication?

Firefighters and emergency medical services (EMS) personnel are usually the first on the scene, responsible for rescuing trapped individuals.

What are some key hazards first responders face during extrication?

Common hazards include traffic, fire, fuel leaks, and vehicle instability.

How do first responders ensure safety while working in high-traffic accident zones?

They establish secure perimeters using cones, flares, and barricades and may coordinate with law enforcement for traffic control.

Why do modern vehicles pose challenges for traditional extrication methods?

Advances in technology, such as high-strength metals, electronic systems, and alternative power sources, require specialized extrication techniques.

What type of metal is commonly used in modern car frames that challenges conventional cutting tools?

Ultra-high-strength steel (UHSS) is frequently used and may require specialized hydraulic cutters.

What are some risks associated with extricating victims from electric or hybrid vehicles?

Risks include electrocution and battery-related hazards, requiring responders to safely disable electrical systems before proceeding.

Why is continuous training necessary for vehicle extrication?

As vehicle designs evolve, first responders must update their knowledge and practice new techniques to ensure safe and effective extrications.

What are the primary concerns when extricating victims from frontal collisions?

Dashboard intrusion and steering wheel compression injuries may require dash rolls or hydraulic spreaders.

Why are side-impact (T-bone) collisions particularly dangerous?

These crashes often cause severe structural deformation, making victim removal more difficult.

How do high-speed crashes complicate vehicle extrication?

They result in extreme vehicle damage, increased debris hazards, and more severe injuries that require immediate medical attention.

What are some special considerations for extricating victims from large vehicles like trucks or buses?

Factors include the vehicle's height, weight, potential hazardous cargo, and the need for heavy-duty equipment such as cranes or jacks.

What is the first step in scene size-up during an extrication operation?

Establishing a safe zone by controlling traffic and securing the area.

Why is a 360-degree assessment of the accident scene important?

It helps responders identify victim entrapment, vehicle stability, and potential hazards before beginning extrication.

What is the role of an incident command center during vehicle extrication?

It coordinates communication, assigns roles, and ensures all emergency teams work efficiently.

What secondary hazards should responders be aware of at a crash scene?

Power lines, fuel leaks, hazardous materials, and unstable vehicles.

Why is vehicle stabilization crucial before extrication begins?

It prevents secondary movements that could further injure victims or endanger responders.

What are some basic vehicle stabilization methods?

Chocking wheels, securing suspension systems, and using cribbing blocks.

How do rescue struts assist in vehicle stabilization?

They provide adjustable support to secure unstable vehicles, preventing unwanted shifts.

What unique challenges do water rescues present in vehicle extrication?

Buoyancy and shifting currents can alter the vehicle's position, requiring special flotation and anchoring techniques.

Chapter 7: Wilderness and Remote Rescues: Survival Tactics and Medical Aid

Navigating the complexities of wilderness and remote rescues is an intricate task that requires a blend of specialized skills and strategic thinking. In these environments, every decision counts, as rescuers face unpredictable challenges that test their expertise and adaptability. Unlike urban settings, where resources are within easy reach, wilderness rescues demand unique problem-solving abilities. Rescuers must possess not only the ability to deliver immediate assistance but also the foresight to anticipate unexpected obstacles that could arise in such unpredictable conditions.

This chapter delves into the essential tactics and skills required for effective wilderness and remote rescues. The discussion encompasses the distinct differences between urban and remote rescue operations, emphasizing the significant role accessibility plays in rescue strategies. Weather and terrain further compound these challenges, making it imperative for search and rescue (SAR) teams to be well-prepared with diverse methodologies tailored for harsh environments. From advanced navigation and survival techniques to improvising medical aid with limited resources, each section provides comprehensive insights into the nuanced processes involved. Additionally, the chapter sheds light on the cultural aspect of SAR and the importance of collaboration across various disciplines to enhance overall response capabilities. Through this exploration, readers will gain a clearer understanding of the multifaceted nature of wilderness rescues and the continuous need for skill development among those committed to saving lives in remote locations.

Overview of Wilderness and Remote Rescues

Rescues in wilderness and remote locations present unique challenges that set them apart from urban environments. Understanding these distinct features is essential for anyone involved in such operations. Wilderness environments are characterized by their vastness, unpredictability, and often harsh conditions. Unlike urban settings where infrastructure and resources are readily available, wilderness areas offer limited access to medical facilities, communication networks, and rescue personnel.

One of the primary differences between urban and remote rescue efforts is accessibility. In urban areas, emergency services can usually reach a location quickly due to established roads and transport systems. However, in remote rescues, the journey to the site can be as challenging as the rescue itself, requiring navigation through forests, mountains, or deserts with few or no paths. Resources such as helicopters or ATVs may be necessary to cover great distances, but even then, weather conditions or difficult terrain can impede their use.

Weather and terrain pose significant environmental challenges in wilderness rescues. Extreme weather conditions like blizzards, heavy rain, or scorching heat can impact both the rescue team and those needing assistance. For example, storms can increase the risk of injury or hypothermia, while extreme heat can lead to dehydration and heatstroke. Additionally, the terrain—whether it's rocky cliffs, dense forests, or steep ravines—can complicate efforts to reach stranded individuals, necessitating specialized equipment and skills to navigate safely.

Search and Rescue (SAR) teams play a critical role in these operations. These teams are comprised of both professional rescuers and trained volunteers who bring expertise in diverse environments and conditions. SAR teams must be well-prepared, physically fit, and knowledgeable about the specific terrains they might encounter. Their training often includes techniques for navigation, survival, first aid, and using technology to locate missing persons swiftly and efficiently.

The distinction in operations becomes evident when comparing the methodologies used in urban versus remote rescues. In urban contexts, the focus might lie more on technological coordination and infrastructure leverage. In contrast, remote rescues often require a more hands-on approach due to the lack of immediate technological aid or structural support systems. The necessity for self-reliance in wilderness scenarios demands that SAR teams be proficient in improvising solutions when faced with unexpected obstacles.

The rising number of individuals engaging in wilderness activities highlights the need for increased awareness and preventive measures. As seen in studies of regions like the Veneto Prealps in Italy, there has been a marked increase in rescue missions tied to outdoor recreation without corresponding improvements in public knowledge of the risks involved (Ciesa et al., 2015). This underscores the importance of public education regarding the inherent dangers of wilderness environments and the preparation needed before embarking on such adventures.

Moreover, SAR operations aren't just about locating individuals; ensuring their safety until they can be evacuated is equally crucial. This involves the ability to administer medical aid with limited resources. SAR teams must be adept at performing primary assessments and stabilizing patients under challenging conditions, often with improvised means due to the scarcity of standard medical supplies.

The cultural aspect of SAR should not be overlooked. It is common for wilderness SAR teams to assist in urban areas when required, utilizing their expertise in expansive searches. However, the primary focus remains on remote rescues where their skills are most in demand (UFO Themes, 2018). Collaboration between different SAR disciplines enhances the effectiveness of rescue missions and strengthens the overall response capacity during emergencies.

Incident Assessment and Safety Measures

In the face of emergencies in remote and wilderness areas, a structured approach is essential for rescuers to effectively evaluate situations and ensure the safety of both victims and themselves. First responders are often required to make rapid assessments under challenging conditions where conventional medical or logistical support may be delayed. A critical first step is to assess the number of victims involved and determine the severity of their injuries. This information acts as the foundation for prioritizing rescue actions and allocating resources.

In such scenarios, triage becomes indispensable. This process involves categorizing victims based on the immediacy of their medical needs. In many cases, life-threatening conditions like severe bleeding or compromised airways must be dealt with promptly. Providing immediate basic care can mean the difference between life and death until more advanced medical help is available (Colling & York, 2010). Field assessments by rescuers, especially those trained in wilderness medicine, are crucial in this stage to stabilize victims and prepare them for eventual transport.

Apart from the initial victim assessment, gauging potential threats is equally important for ensuring overall safety during a rescue operation. For instance, the unpredictability of avalanches or sudden weather shifts demands that rescuers stay vigilant and informed about local environmental conditions. Weather can change dramatically, turning what seems like a straightforward mission into a perilous endeavor. By closely monitoring weather forecasts and maintaining a constant lookout for signs of impending risks, rescuers can adjust plans accordingly. They may even need to enact preventive measures to avoid secondary incidents like hypothermia or dehydration during protracted operations.

Navigating through difficult terrains poses another significant challenge. Landscapes that include steep cliffs, rocky slopes, or dense forests require special attention as they can increase the risk of injury to both rescuers and victims. Hazards such as loose rocks, slippery trails, or concealed crevices necessitate slow and careful movement, often demanding specialized equipment and trained personnel to manage these risks effectively. Moreover, encounters with wildlife add another layer of complexity, especially when dangerous animals might be present. Understanding the behavior of local fauna and implementing strategic measures to keep both the victims and rescue teams safe is vital in these settings.

Communication is another cornerstone of successful emergency management in remote settings. Reliable communication allows for effective coordination, reassures victims, and facilitates timely updates on the situation's developments. Due to limited infrastructure in wilderness areas, traditional mobile networks are often unreliable, making satellite phones and radios critical tools. These devices help maintain contact with command centers and allow for the swift relay of information regarding evolving challenges, changes in victim status, or resource requirements. Consistent communication also supports collaborative efforts involving multiple agencies or volunteers, enhancing the operational capacity and precision of rescue missions (Wozniakowski-Zehenter, 2024).

In practice, communication strategies should begin before field deployment. Pre-mission briefings outline expected risks and specify the roles and responsibilities of team members. Once on site, regular check-ins using radios or satellite-based systems confirm ongoing situational awareness and provide updates on rescue progress. Redundant communication lines are often established to compensate for any device failures or signal obstructions due to terrain features. For extended operations, portable power sources may be utilized to ensure communication devices remain functional.

Search and Rescue Techniques

Wilderness and remote rescues present unique challenges that necessitate specialized strategies for locating and rescuing individuals. These operations often require a blend of traditional methods with advanced technologies to effectively address the vast and often treacherous landscapes involved.

One fundamental strategy in wilderness search and rescue (SAR) operations is the use of grid searches and tracking techniques. Grid searches involve dividing the search area into small, manageable sections, allowing searchers to systematically cover every part. This meticulous approach ensures no area is overlooked, which is crucial when time is of the essence. Trained personnel utilize tracking techniques by observing signs such as footprints, broken branches, or disturbed vegetation, which can indicate the direction a person may have traveled. This method, while traditional, remains effective in guiding teams towards potential locations of missing persons. The combination of grid searches and sign tracking maximizes the chances of locating an individual promptly, which can be critical for survival, particularly in extreme weather conditions or challenging terrains (vcsar1admin, 2017).

Helicopters play a pivotal role in aerial reconnaissance and victim extraction. In areas where ground access is limited or impossible, helicopters provide a bird's-eye view, enabling teams to quickly scan large swaths of land and identify possible clues from above. Equipped with advanced imaging technology, including thermal cameras, helicopters are invaluable for spotting individuals who may be hidden under dense foliage or snow. Once a victim is located, helicopters facilitate rapid extraction, reducing the time it takes to transport them to safety or medical care. However, helicopter operations are heavily dependent on weather conditions and require careful coordination with ground teams to ensure operational safety and effectiveness (Vincent-Lambert et al., 2023).

Water-based operations introduce additional complexities but are just as important in specific rescue scenarios. Drones and specialized equipment are increasingly used for these operations, especially in areas like rivers, lakes, or coastal regions. Unmanned Aerial Vehicles (UAVs), commonly known as drones, can be deployed for aerial reconnaissance over water, significantly enhancing visibility and reach without placing human rescuers at risk. Equipped with cameras and other sensors, these drones can capture real-time data and relay information back to the command center, assisting in strategic planning. Specialized equipment, such as underwater drones, can also be utilized to explore submerged areas or assist in rescues involving boating accidents. The deployment of UAVs has been shown to mitigate risks associated with water-based rescues, making operations faster and safer by avoiding unnecessary exposure of rescuers to potentially hazardous environments (Vincent-Lambert et al., 2023).

Coordinated rescue efforts with air support form another crucial aspect of successful wilderness SAR missions. Effective coordination enhances communication between different teams—ground personnel, air operators, and any supporting units—and ensures resources are used efficiently. Air support not only facilitates quicker searches and rescues but also improves overall situational awareness. For instance, radio communication restored through UAVs can create a reliable link between scattered team members in terrain where traditional communication lines might fail due to distance or obstacles. This kind of coordination reduces response times, ensures swift movement of resources, and helps prioritize rescue operations based on updated information from all sources involved. Establishing clear communication protocols and having contingency plans are essential guidelines for teams managing these complex operations (Vincent-Lambert et al., 2023).

Grid searches remain foundational, utilizing trained personnel and technology to systematically comb areas and enhance detection capabilities. For aerial reconnaissance, employing helicopters equipped with advanced imaging tools allows swift scanning of vast areas inaccessible by foot. Water-based rescues, facilitated by drones and specialized gear, leverage technology to navigate aquatic landscapes safely. Central to these operations is robust coordination among diverse teams, integrated with air support to synchronize efforts efficiently.

Medical Aid and Patient Management

In the vast expanse of wilderness and remote areas, administering medical care requires not only a keen understanding of emergency response but also adaptability and resourcefulness. First responders and healthcare providers venturing into these environments face unique obstacles that demand specialized knowledge and skills for effective medical intervention.

Conducting a primary assessment in the wilderness is an essential initial step, focusing on vital signs to gauge the urgency of the situation. The ABCs—Airway, Breathing, and Circulation—serve as the foundation for evaluating a patient's condition. Ensuring that a patient's airway is clear and unobstructed is paramount. This could involve positioning techniques or using makeshift tools to maintain patency if standard equipment is unavailable (Pittman et al., 2020). Evaluating breathing may require listening to breath sounds without the aid of stethoscopes, while checking circulation involves identifying signs of shock or severe blood loss, often without the luxury of extensive diagnostic tools.

Wilderness environments frequently expose individuals to traumatic injuries such as fractures and sprains. Treating these injuries effectively with limited resources involves improvisation and creativity. Splinting a fracture, for example, might require using branches, clothing, or other natural elements available in the environment. Creating makeshift support for sprained limbs can help immobilize the injury and prevent further harm until professional medical help is available. Additionally, understanding how to manage bleeding with improvised tourniquets is crucial, as this can be lifesaving when resources are scarce (Hawkins & Simon, 2021).

Medical emergencies like hypothermia pose significant threats in remote settings. Recognizing early signs, such as shivering and confusion, and responding promptly by providing insulation, body warmth, and shelter can stabilize a patient's condition. In hypoallergenic scenarios, such as anaphylactic shock, swift action is crucial despite potentially having limited access to medications. If epinephrine is unavailable, managing symptoms through position changes and efforts to maintain comfort becomes critical while awaiting evacuation (Peixotto et al., 2021).

Preparation plays a significant role in effective wilderness medicine. A well-equipped field medical kit is indispensable. Basic supplies like bandages, antiseptic wipes, and pain relievers form the core of immediate care, but so does innovation in utilizing local resources when kits run dry. For instance, clean water sources can be used for wound irrigation, while herbal plants known for their medicinal properties may offer temporary relief or treatment. Creating such kits requires foresight and a thorough understanding of both necessary supplies and potential wilderness substitutes (Pittman et al., 2020).

Developing guidelines for equipping field medical kits includes ensuring redundancy in critical items like bandages or gauze, which can be repurposed if needed. Practitioners should undergo training that emphasizes adaptability, learning how to efficiently use and replace these essentials in resource-constrained environments. While high-tech solutions are beneficial, basic survival tactics and common sense remain valuable assets for effective remote medical care.

Ultimately, the ability to provide competent medical assistance in the wilderness hinges on the balance between knowledge and adaptability. Training programs focusing on practicing medicine under austere conditions are invaluable, preparing first responders to address not only predictable injuries but also the unexpected challenges presented by nature (Hawkins & Simon, 2021). Using evidence-based methodologies tailored to the unpredictability of nature fosters resilience among those tasked with delivering care in isolation.

Essential Navigation and Survival Skills

Navigating the challenging terrains of wilderness and remote areas requires a thorough understanding of land navigation tools such as maps, compasses, and GPS devices. For rescuers and victims alike, mastering these tools is essential for effective orientation and safety in unfamiliar landscapes. Maps offer a detailed depiction of the area, showing key features like rivers, mountains, roads, and trails. Understanding topographic symbols and scales on a map is crucial for route planning and identifying potential hazards. Compasses provide direction based on magnetic north, allowing users to orient maps accurately and navigate even when visibility is poor or landmarks are not visible. The use of a compass can be complemented by GPS devices, which utilize satellite signals to pinpoint exact locations, offering precision that traditional methods may lack. However, reliance solely on GPS can be risky due to potential signal loss in dense forests or mountainous terrain, highlighting the importance of proficiency in both traditional and modern navigation techniques (Marsteiner, 2024).

In survival situations, prioritizing basic needs such as shelter, water, fire, and food is fundamental. Shelter provides protection from harsh weather conditions, helping prevent hypothermia in cold environments or heatstroke in hot climates. Constructing emergency shelters using available materials like tarps, branches, and leaves creates a safe haven. A tarp can serve as a rainfly or windbreak, and natural elements can be used to improve insulation. Lean-tos, debris huts, and tarp shelters are common structures that vary in complexity but all aim to offer warmth, dryness, and security.

Water is another critical survival need, as it is vital for hydration, digestion, and overall bodily functions. Identifying water sources such as streams, lakes, or even dew collected from leaves can be life-saving. Once located, it is necessary to purify the water to make it safe for consumption. Boiling is the most reliable method, killing most pathogens with just a few minutes of bubbling. If boiling isn't possible, water purification tablets offer an alternative, though they may impart an unpleasant taste. Filters are also effective, especially in removing sediment and some bacteria, but should be chosen carefully based on their filtration capabilities. In extreme cases, even sunlight, which utilizes UV radiation to kill pathogens when exposed long enough, can be considered for disinfection (Survivor: The Ultimate Guide to Thriving in Extreme Situations, 2025).

Fire serves multiple functions, providing warmth, light, and the ability to cook food, making it a priority in many survival scenarios. Learning how to start a fire without matches is beneficial for emergency preparedness. Methods such as using flint and steel, ferro rods, or friction-based tools like bow drills are crucial skills. Collecting appropriate tinder, kindling, and fuel ensures success in igniting and sustaining a fire. Always assemble your fire in a controlled environment, clear of flammable materials, and ensure you have sufficient ventilation if constructing near or in a shelter. Practicing fire-building techniques beforehand can significantly improve your chances of successfully creating and maintaining a fire in an urgent situation (Marsteiner, 2024).

Food procurement becomes critical when stuck in remote areas for extended periods. While our bodies can survive longer without food than without water, energy depletion will severely impact physical and cognitive faculties over time. Wild edibles, such as berries, nuts, and edible plants, provide necessary nutrients. Familiarity with local flora and fauna aids in selecting safe options while avoiding toxic counterparts. Additionally, fishing and trapping can secure more substantial protein sources. Simple traps like snares can catch small game, while improvised fishing lines and hooks make obtaining fish feasible.

Finally, signaling devices can dramatically increase the likelihood of being found by rescuers. Mirrors are among the most effective tools, capable of reflecting sunlight over great distances to attract attention. Alternately, whistles and auditory signals complement visual ones, especially in times of low visibility, ensuring that search parties are aware of your location. Preparing and practicing the use of these items should form part of any comprehensive survival plan.

Bringing It All Together

In this chapter, we delved into the distinctive challenges and necessary strategies involved in wilderness and remote rescues. These operations differ significantly from urban rescues due to factors such as limited accessibility, unpredictable weather, and difficult terrains. Understanding these elements is crucial for those involved in rescue efforts, including Search and Rescue (SAR) teams that consist of both professional rescuers and volunteers. The need for heightened awareness and preparation was also emphasized, as an increasing number of people engage in outdoor activities without fully understanding the risks. From assessing victim conditions to coordinating complex operations with air support, it is clear that a structured approach is paramount to success.

The essential skills required for these rescue missions extend beyond physical endurance and technical expertise. Effective communication, navigation, and medical intervention under resource constraints are vital components of a successful operation. By focusing on self-reliance and improvisation, SAR teams can adapt to unexpected obstacles and provide critical care to victims until more comprehensive help arrives. Through the use of technology combined with traditional methods, rescuers can enhance their efficiency while ensuring safety in challenging environments. This chapter highlights the importance of traversing these remote landscapes with preparedness, fostering a proactive mindset among first responders and emergency personnel.

Chapter 7 Test Questions and Answers

What are some key challenges that set wilderness rescues apart from urban rescues?

Wilderness rescues are characterized by limited access to medical facilities, difficult terrain, unpredictable weather, and the need for specialized navigation and survival skills.

How does accessibility differ between urban and remote rescue operations?

Urban rescues benefit from established roads and infrastructure, allowing for quick response times. Remote rescues require travel through forests, mountains, or deserts, often needing ATVs, helicopters, or foot navigation.

Why is terrain a major challenge in wilderness rescues?

Steep cliffs, dense forests, and rocky slopes make access difficult, increasing the risk of injury for both rescuers and victims while requiring specialized equipment and training.

How does weather impact search and rescue (SAR) operations?

Extreme weather, such as blizzards, heatwaves, and storms, can hinder visibility, create additional hazards, and affect both rescuers and victims, increasing the risk of hypothermia or dehydration.

What role do SAR teams play in wilderness rescues?

SAR teams, made up of professionals and trained volunteers, specialize in search techniques, navigation, first aid, and survival skills to locate and stabilize victims before evacuation.

How do search techniques differ in wilderness rescues compared to urban settings?

Wilderness searches rely on grid searches, tracking footprints and environmental clues, while urban rescues leverage surveillance systems, GPS tracking, and emergency response infrastructure.

What is triage, and why is it important in remote rescues?

Triage is the process of categorizing victims based on medical urgency to prioritize treatment, ensuring life-threatening conditions like severe bleeding or airway blockages are addressed first.

What factors must rescuers assess when arriving at a remote rescue site?

Rescuers must evaluate the number of victims, severity of injuries, environmental hazards, weather conditions, and available resources before determining the best rescue strategy.

How do SAR teams maintain communication in remote areas?

Traditional mobile networks may not work, so teams rely on satellite phones, radios, and UAVs to establish and maintain communication with command centers.

What is the significance of pre-mission briefings in SAR operations?

Pre-mission briefings outline risks, roles, communication strategies, and contingency plans, ensuring all team members are prepared for potential challenges.

How are helicopters used in wilderness rescues?

Helicopters provide aerial reconnaissance, locate victims using thermal imaging, and enable rapid extraction from inaccessible locations.

What role do drones play in SAR missions?

Drones are used for aerial searches, mapping terrain, relaying real-time data, and even conducting water-based rescues with specialized sensors.

What medical emergencies are common in wilderness rescues?

Hypothermia, heatstroke, fractures, dehydration, severe bleeding, and anaphylactic reactions are common due to environmental exposure and injuries from falls.

How do rescuers stabilize patients with limited medical supplies?

They use improvised splints, makeshift tourniquets, natural insulation for hypothermia, and rely on creative problem-solving to manage injuries until evacuation.

Why is navigation training essential for SAR teams?

Wilderness rescues require expertise in using GPS, maps, compasses, and natural landmarks to locate victims in remote areas without established paths.

What are the benefits of coordinated air support in SAR missions?

Air support speeds up search efforts, improves situational awareness, and ensures efficient resource allocation for ground teams.

How can preventive measures reduce the need for wilderness rescues?

Public education on survival skills, proper equipment, risk awareness, and emergency preparedness can help prevent accidents and reduce rescue incidents.

Why must SAR teams be physically fit?

Rescue operations require endurance, strength, and agility to navigate harsh terrain, carry equipment, and assist victims in physically demanding conditions.

What are some common hazards SAR teams face during remote rescues?

Loose rocks, avalanche risks, wildlife encounters, rapidly changing weather, and unstable terrain can all pose dangers during operations.

How do SAR teams manage extended rescue operations?

They establish temporary shelters, ration supplies, rotate team members for endurance, and use portable power sources for communication devices to sustain long-term operations.

Chapter 8: Mass Casualty Incidents: Coordination, Triage, and Crisis Management

Managing mass casualty incidents presents a complex challenge that calls for a well-coordinated response, efficient systems of triage, and effective crisis management strategies. These situations overwhelm local resources and demand rapid response from emergency personnel who must prioritize care under immense pressure. Understanding the scope and urgency inherent in these incidents is vital for anyone involved in emergency response, whether they are seasoned first responders, firefighters, or students preparing to enter the field. Each incident is unique, presenting different challenges based on its nature and scale, which makes adaptability and preparedness crucial.

The chapter delves into important aspects of mass casualty management, beginning with the coordination mechanisms that unify various agencies in their response efforts. It explores the safety measures that protect both responders and victims while maintaining an organized scene under stressful conditions. The intricacies of triage systems are examined, highlighting how they help allocate scarce resources efficiently. Transportation logistics, another critical component, are addressed to demonstrate the importance of swift and proper patient relocation. Furthermore, the chapter touches on specialized considerations required to improve response efficacy, clarifying the nuances of each type of incident. By the end, readers will have a comprehensive understanding of how these elements collectively enhance the ability to manage mass casualties effectively.

Understanding Mass Casualty Incidents

Mass casualty incidents (MCIs) present a unique set of challenges and complexities due to the sheer volume of casualties that surpass local healthcare resources, creating circumstances far beyond what routine multi-patient incidents entail. In these situations, local healthcare systems often find themselves strained, struggling to provide adequate care not only for those directly affected by the MCI but also for patients requiring usual day-to-day medical services. Hospitals need to have surge plans in place to prepare for such events because they can rapidly exhaust available resources. These plans should account for various scenarios, including events with gradual signs and sudden-onset disasters.

The nature of MCIs can vary significantly based on the type of incident. Natural disasters like earthquakes, hurricanes, and floods often result in MCIs due to their potential to impact large geographic areas and populations. Similarly, terrorist events, particularly those involving explosives or chemical agents, can lead to high numbers of casualties with complex injury patterns. Transportation accidents, whether involving planes, trains, or automobiles, frequently cause mass casualties due to the number of individuals typically involved in these incidents. Industrial mishaps, such as chemical spills or factory explosions, represent another category of MCIs, posing additional hazards related to toxic exposure risks. Public health emergencies, such as pandemics or widespread foodborne illness outbreaks, constitute another form, straining healthcare systems over extended periods (DeNolf & Kahwaji, 2019).

One of the most pressing challenges during MCIs is the overwhelming burden placed on medical personnel. Emergency responders, doctors, nurses, and support staff must manage an influx of patients needing immediate attention while working with limited resources. This situation demands rapid decision-making and prioritization skills, often under high-pressure conditions. The psychological toll on responders can be substantial, as they face the stress of potentially life-and-death choices, fatigue from long hours, and emotional strain from witnessing significant trauma (Clarkson & Williams, 2023).

Resource scarcity further compounds the difficulties inherent in MCIs. Medical supplies, such as bandages, medications, and lifesaving equipment, can quickly become depleted. Additionally, hospitals and clinics may lack the physical space necessary to accommodate all patients, necessitating triage systems to prioritize care. Effective triage becomes essential; systems like START (Simple Triage and Rapid Treatment) help sort patients based on urgency, enabling efficient allocation of scarce resources. However, resource scarcity goes beyond medical supplies and facilities; it also includes human resources. During an MCI, there may be insufficient trained personnel available to manage the scene and provide treatment, highlighting the importance of community preparedness and cross-agency coordination.

The psychological impact of MCIs is profound, affecting both responders and survivors. Responders experience heightened stress and pressure, needing coping mechanisms to handle trauma exposure. Survivors grapple with physical injuries and emotional distress from experiencing a traumatic event. Critical Incident Stress Management programs play a crucial role in supporting mental well-being for everyone involved in responding to or affected by an MCI. These programs provide counseling and debriefing sessions to process experiences and offer psychological first aid to mitigate longer-term effects.

Natural disasters, such as earthquakes, hurricanes, and floods, are among the most common types of MCIs, as they affect vast areas and large populations simultaneously. These events can severely damage infrastructure, making access to affected regions challenging for emergency responders. The unpredictability of natural disasters complicates response efforts further and underscores the importance of preparation and resilience for emergency management agencies.

Terrorist attacks, particularly those using explosives or chemical weapons, require specialized response strategies due to their unpredictable nature and potential for causing wide-scale panic. These events necessitate rapid identification of safe zones and effective communication to coordinate response efforts across multiple agencies. Transportation accidents, including collisions involving buses, trains, or airplanes, often occur without warning and can result in numerous casualties within a concentrated area. These incidents demand quick mobilization of nearby medical resources and coordinated triage efforts at the scene to expedite patient treatment and transfer.

Industrial mishaps, such as chemical plant explosions or toxic leaks, pose significant challenges due to the potential hazards associated with the substances involved. Responders must wear appropriate protective equipment and follow stringent safety protocols to prevent further harm. Furthermore, public health emergencies, like pandemics, stretch healthcare systems over extended periods with continuous demands for resources and personnel. These events highlight the necessity of robust planning and coordination among governmental and non-governmental organizations to sustain prolonged response efforts.

Command and Coordination in MCI

The Incident Command System (ICS) plays a pivotal role in managing mass casualty incidents by providing a structured framework that defines key roles such as the Incident Commander and Operations Officer. These roles are essential for ensuring that response efforts are coordinated efficiently, with the Incident Commander holding the overall responsibility for incident management while the Operations Officer directs actions to achieve the set objectives (2014). This delineation of responsibilities fosters an organized approach, allowing for clarity and focus during high-pressure situations.

A cornerstone of effective incident management is the establishment of unified command, which facilitates seamless coordination among various agencies involved in emergency response. This includes fire services, Emergency Medical Services (EMS), police, emergency management, hospitals, and federal agencies such as the Federal Emergency Management Agency (FEMA). Unified command allows these diverse organizations to work collaboratively without compromising their individual authorities. By combining expertise and resources, agencies can tackle complex incidents more effectively, leveraging each other's strengths to mount a comprehensive response effort (FEMA, 2024).

Communication is another critical component of successful incident command and agency coordination. Effective communication protocols ensure interoperability between agencies through the use of mobile command centers and standardized radio frequencies. This interoperability is crucial for maintaining clear lines of communication across all participating entities, enabling them to share real-time information and updates throughout the incident's progression. A common communications plan, supported by tools like the ICS 205 form, helps ensure that everyone involved is on the same page, reducing the likelihood of miscommunications that could hinder response operations (2014).

Guideline: It is imperative that all personnel are familiar with integrated communication systems and adequately trained to operate them under pressure. Regular drills and exercises help reinforce these skills, ensuring that during an actual incident, communication flows smoothly across agencies and jurisdictions.

In situations where additional resources are required, mutually coordinated efforts come into play. Coordination requests for additional personnel and equipment help optimize regional healthcare support, ensuring that sufficient resources are available to meet the demands of the incident. This process involves strategic planning and collaboration between local, state, and federal agencies, often facilitated through pre-existing mutual aid agreements. Such agreements enable rapid mobilization of resources from neighboring regions, alleviating strain on local responders and infrastructure (FEMA, 2024).

Guideline: Establishing pre-incident agreements and understanding the protocols for resource allocation are vital for swift deployment during emergencies. Agencies should routinely review and update these agreements to reflect current capabilities and limitations.

Overall, the role of incident command and agency coordination in managing mass casualty incidents cannot be overstated. The ICS provides the necessary structure for defining roles and responsibilities, ensuring that all entities involved work towards common objectives. Unified command enhances this coordination by bringing together various agencies to collaborate efficiently, each contributing its unique capabilities to the response effort. Effective communication protocols further streamline this process, facilitating real-time information sharing and operational cohesion. Ultimately, through mutually coordinated efforts, agencies can maximize resource availability, improving overall response efficacy and outcomes in mass casualty scenarios.

Scene Safety and Patient Categorization

In the realm of mass casualty incidents (MCIs), ensuring scene safety and efficient triage is critical to enhancing response efficacy. This involves a calculated approach that begins with a comprehensive initial scene size-up. The primary goal here is to evaluate the extent of damage and proactively identify any potential secondary threats. These threats may include hazardous materials or active assailants, which can compromise responder safety and complicate rescue efforts. For instance, in situations involving chemical spills or toxic substances, responders must wear appropriate personal protective equipment (PPE) to mitigate exposure risks. Similarly, if there is an active shooter scenario, securing the area becomes paramount to prevent further casualties.

Once the scene is assessed, establishing perimeter zones is essential for organizing response activities and maintaining safety. Different zones, such as hot, warm, and cold, are delineated based on the level of danger present. The hot zone is closest to the incident and requires responders to don specialized PPE due to the high-risk environment. Activities within this zone are generally confined to lifesaving interventions and stabilization procedures. In contrast, the cold zone is situated away from immediate hazards, serving as a logistics hub where planning, support operations, and coordination take place. By clearly defining these zones, responders can focus on their tasks without unnecessary overlap, reducing confusion and improving response efficiency.

A cornerstone of MCI management is triage, which involves prioritizing patients based on the severity of their injuries to maximize survival rates. Various systems have been developed to facilitate effective triage. For instance, START (Simple Triage and Rapid Treatment) and JumpSTART are widely used methodologies that help categorize patients quickly. START is typically employed for adults, utilizing observable criteria to assign priority levels such as immediate, delayed, or minor. On the other hand, JumpSTART is tailored for pediatric patients, acknowledging the physiological differences in assessing younger victims. SALT (Sort, Assess, Life-saving interventions, Treatment/Transport) provides a structured approach that is adaptable to various scenarios, allowing for flexibility depending on resource availability and the specific nature of injuries encountered.

Triage tags play a vital role in patient tracking and management throughout the incident response. These tags, attached to each patient, provide crucial information about the individual's condition, treatment priorities, and corresponding triage category. Information on the tags is often color-coded for clarity, enabling quick visual identification of needs. Additionally, tags facilitate communication among responders by providing a snapshot of the patient's status, thereby streamlining medical care and transport decisions. As conditions evolve and resources fluctuate, reassessment is necessary to ensure that each patient receives the most appropriate level of care. Adjustments to triage categories reflect changes in patient stability and resource allocation, thus maintaining a dynamic and responsive treatment strategy.

While understanding the frameworks guiding scene safety and triage is pivotal, it is equally important to consider guidelines that enhance responder safety during operations. Those engaged in the initial scene size-up should systematically assess the environment, taking note of visible hazards while also considering potential unseen dangers, such as structural instabilities or toxic gas emissions. For example, in the aftermath of a building collapse, structural engineers might be consulted to evaluate the integrity of remaining structures, while hazardous materials teams can test air quality to determine if respiratory protection is needed.

Similarly, establishing perimeter zones requires strategic placement based on the nature of the incident and existing threats. Guidelines recommend that hot zones encompass areas immediately surrounding the epicenter of the hazard, necessitating the use of protective gear to access safely. Warm zones serve as decontamination areas where individuals leaving the hot zone undergo processing before entry into the cold zone. Effective perimeter management not only protects personnel but also ensures a systematic flow of operations, minimizing bottlenecks and preventing unauthorized access to dangerous areas.

In practice, ensuring responder safety extends beyond physical protection to encompass mental well-being, with teams trained to recognize signs of stress and fatigue that could impair judgment or response ability. Given the intensity and unpredictability inherent in MCIs, responders must be equipped with strategies for stress management and psychological resilience. Training programs typically cover recognition of stress symptoms, techniques for maintaining composure under pressure, and protocols for peer support, all of which contribute to a safer, more resilient workforce.

Field Treatment and Patient Transport

In the realm of mass casualty incidents (MCIs), effective field treatment methods and transport logistics are crucial to saving lives. The primary objective during these crises is to execute life-saving interventions promptly, ensuring victims receive immediate attention and support. Key interventions in such scenarios include airway management, bleeding control, and fracture stabilization, each central to preserving life in the chaotic environments typical of MCIs.

Airway management involves clearing and maintaining open airways to ensure oxygen can be effectively exchanged. This may require techniques such as the head-tilt-chin-lift maneuver or, in more severe cases, endotracheal intubation. Bleeding control is another critical intervention, as uncontrolled hemorrhage is one of the leading causes of preventable death. Techniques such as direct pressure application, tourniquet use, and hemostatic dressing application are essential skills first responders must employ quickly. Fracture stabilization prevents further injury and reduces pain, often employing splints or other immobilization devices to secure broken limbs.

Setting up field treatment areas is a strategic endeavor aimed at optimizing available resources across different care levels. These areas are designated spaces, often categorized into zones based on injury severity, allowing for efficient patient assessment and treatment. By organizing areas according to resource needs, responders can administer tailored care efficiently, improving patient outcomes. Equipment such as portable medical kits, stretchers, and communication tools are strategically positioned within these zones to facilitate seamless operations.

Ambulance staging plays a pivotal role in transport logistics, preventing congestion at the scene and prioritizing the evacuation of critically injured patients. Proper staging ensures ambulances are not only ready for immediate deployment but also coordinated with hospital capacities to avoid overwhelming single facilities. EMS personnel manage resources by categorizing patients based on urgency, aligning transportation efforts with available regional medical facilities.

Pre-notification of receiving hospitals is a critical component within transport logistics. Timely communication enables hospitals to prepare for incoming patients, ensuring capacity planning aligns with real-time demands. This process aids hospitals in rapidly adjusting their surge capacity, whether that involves converting waiting rooms into additional treatment spaces or mobilizing extra staff. Pre-notification also facilitates balanced distribution of patients across multiple facilities, minimizing the strain on any single hospital and enhancing the overall effectiveness of the response effort.

Triage systems and methods form the backbone of decision-making processes during MCIs, guiding first responders in determining treatment priorities. Systems like START (Simple Triage and Rapid Treatment) and SALT (Sort, Assess, Life-Saving Interventions, Triage/Treatment) provide structured, standardized approaches for categorizing victims based on injury severity and resource availability. Effective triage ensures the most severely injured receive priority care, maximizing survival rates.

Guidelines for applying triage systems underscore the necessity of clear protocols and regular training. Responders must be well-versed in these systems, executing them swiftly under pressure. Incorporating triage tags and patient tracking ensures ongoing monitoring and reassessment, accommodating shifts in patient conditions or resource availability. Notably, guidelines emphasize adapting to dynamic circumstances, encouraging improvisation where standard procedures fall short.

In practice, life-saving interventions in the field demand rapid identification and action. First responders, equipped with foundational medical skills, must act decisively, balancing protocol adherence with situational fluidity. Coordinated teamwork amplifies these efforts, as varied expertise contributes to comprehensive patient assessment and care delivery.

Additionally, collaboration extends beyond the incident scene, involving coordination with external agencies and community resources. Such partnerships enhance logistical efficiency, leveraging broader networks for both immediate medical assistance and sustained recovery efforts post-incident.

Moreover, continuous learning from past incidents informs future MCI responses, emphasizing the value of debriefing sessions and outcome evaluations. Lessons learned contribute to refined practices, addressing previously encountered challenges and reinforcing successful strategies.

Ultimately, the success of field treatment methods and transport logistics hinges on preparedness, adaptability, and cohesive action. Ensuring first responders are equipped with the necessary skills, knowledge, and resources fosters resilience against the unpredictable nature of MCIs. Through streamlined operations and collaborative engagement, emergency services can significantly improve response efficacy, safeguarding lives even in the most challenging circumstances.

Special Considerations and Communication During MCIs

In the realm of mass casualty incidents (MCIs), addressing unique event challenges, especially during active shooter situations and explosive events, is paramount for the safety and effectiveness of first responders. Active shooter scenarios demand rapid decision-making under extreme pressure. Tactical triage becomes essential when shots are still being fired. Responders must prioritize immediate life-threatening conditions while ensuring their safety. Rapid extraction techniques, such as using ballistic shields or rescue task force approaches where law enforcement and medical personnel work simultaneously, can minimize exposure to danger while swiftly attending to victims. This collaborative approach provides a critical lifeline in chaotic environments, emphasizing the need for cohesive planning and training.

Explosive events present equally daunting hurdles, requiring special attention to blast injuries which often result in complex trauma including shrapnel wounds, burns, and blast lung. Emergency responders must be cautious of secondary devices—explosive devices strategically placed to target rescuers following an initial event. The presence of these threats mandates thorough sweeps and careful scene management before medical personnel can fully engage. Utilizing bomb squads to clear these threats ensures that subsequent medical interventions proceed without risking further casualties, highlighting the importance of integrated unit coordination.

Effective crisis communications play a pivotal role in managing public perception and expectations during MCIs. Public messaging needs to be timely and precise to prevent misinformation, which can lead to panic or confusion among the population. Establishing official communication channels, such as regular press briefings led by designated spokespersons, helps maintain control over the narrative. Accurate information dissemination not only aids the public but also supports responder operations by reducing unnecessary inquiries and allowing them to focus on critical tasks. In this digital age, social media platforms can serve as powerful tools for real-time updates, provided they are managed by trained communication teams aware of misinformation risks.

The psychological impact of MCIs cannot be overstated. First responders regularly face traumatic scenes, and without appropriate mental health support, they risk long-term psychological harm.

Critical Incident Stress Management (CISM) offers structured support designed to help responders process traumatic experiences. This approach includes pre-incident education, post-incident debriefings, and continued monitoring for signs of stress-related conditions such as PTSD. Implementing CISM protocols ensures that both responders and victims receive necessary psychological care, promoting resilience and recovery within affected communities.

Setting up field treatment areas effectively is crucial to address specific needs during MCIs. These areas provide organized spaces where medical personnel can perform triage, offer immediate care, and prepare patients for transportation. Factors such as proximity to the incident site, accessibility for ambulances, and safety from ongoing hazards determine ideal locations for these setups. Clear guidelines regarding the establishment and operation of these zones enhance overall response efficiency, guiding personnel through a standardized process that maximizes resource utilization and patient outcomes.

Providing psychological first aid extends beyond formal CISM programs, encompassing immediate emotional support delivered directly at the scene. Trained personnel offer comfort and reassurance, helping victims navigate the shock and chaos inherent in disaster situations. This intervention underscores the necessity of integrating mental health strategies within MCI plans, recognizing that emotional wounds often accompany physical injuries.

Determining transport priorities poses another layer of complexity in MCI management. Effective prioritization is vital to ensure that those with the most critical needs reach advanced medical facilities promptly. Triage officers assess patients based on injury severity, stability, and available resources, making crucial decisions about transport order. Coordinating with hospitals to manage patient distribution prevents healthcare facilities from becoming overwhelmed, thereby maintaining treatment standards across the board.

Lastly, special considerations for specific MCIs must be accounted for during planning and execution phases. Each type of incident, whether it involves active shooters, explosives, or natural disasters, presents distinct challenges. Tailoring response plans to accommodate these unique aspects improves overall effectiveness. For example, in incidents involving hazardous materials, responders require specialized equipment and knowledge to safely handle contamination risks, ensuring both their protection and the safety of victims in proximity.

By addressing the unique challenges posed by each event type, optimizing crisis communications, and acknowledging the profound psychological toll on those involved, emergency services can enhance their preparedness and response capabilities. This multi-faceted approach equips responders not only to manage the immediate aftermath of MCIs but also to foster resilience and recovery within the communities they serve. Through meticulous preparation and thoughtful execution of these principles, first responders stand prepared to navigate the complexities of mass casualty incidents with confidence and competence.

Final Thoughts

In this chapter, we have delved into the complexities of managing mass casualty incidents (MCIs), which require meticulous planning and execution to ensure effective response efforts. We explored various coordination mechanisms that underline the necessity for a unified command structure and streamlined communication channels among agencies. Special emphasis was placed on understanding the multifaceted nature of MCIs, whether they arise from natural disasters, terrorist attacks, or industrial accidents. By examining the crucial role of triage systems and transportation logistics, we highlighted how these elements work in concert to maximize efficiency and save lives in high-pressure situations.

As readers, particularly firefighters, first responders, and students, reflect on these insights, it is essential to grasp the importance of preparation and adaptability in facing such unpredictable scenarios. This chapter aimed to equip you with the foundational knowledge needed to navigate and manage each facet of an MCI effectively. By emphasizing scene safety, patient prioritization, and psychological care for responders, we underlined the comprehensive strategies necessary to enhance response efficacy. Ultimately, these pieces form a cohesive response framework that not only addresses immediate needs but also fosters resilience in both responders and affected communities.

Chapter 8 Test Questions and Answers

What is one of the primary challenges emergency responders face during mass casualty incidents (MCIs)?

The overwhelming burden of managing a large number of patients with limited resources.

How does resource scarcity impact MCI response efforts?

Medical supplies, space, and trained personnel may be insufficient, requiring effective triage and coordination.

What system is commonly used to prioritize patient care in MCIs?

The START (Simple Triage and Rapid Treatment) system.

How does psychological stress affect responders during an MCI?

They experience high-pressure decision-making, fatigue, and emotional strain from witnessing trauma.

What support program helps responders cope with the psychological impact of MCIs?

Critical Incident Stress Management (CISM) programs provide counseling and debriefing sessions.

What are some common natural disasters that can lead to MCIs?

Earthquakes, hurricanes, and floods.

Why are terrorist attacks particularly challenging for emergency response?

They often cause wide-scale panic and require specialized response strategies.

What types of transportation accidents can cause MCIs?

Bus, train, and airplane collisions.

Why are industrial accidents, such as chemical spills, difficult to manage in MCIs?

They involve hazardous materials that require protective equipment and specialized response teams.

How do public health emergencies differ from other MCIs?

They place prolonged demands on healthcare systems and require sustained coordination.

What system provides structure in MCI management?

The Incident Command System (ICS).

What is the role of the Incident Commander during an MCI?

To oversee the overall management and ensure coordination among responders.

What is the purpose of unified command in emergency response?

To enable multiple agencies to collaborate effectively without compromising authority.

Why is communication critical during MCIs?

It ensures interoperability between agencies and prevents miscommunication.

How do mutual aid agreements support emergency response?

They allow rapid resource mobilization from neighboring regions.

What are the three response zones in an MCI?

Hot zone (immediate danger), warm zone (decontamination area), and cold zone (safe logistics hub).

Why is triage necessary in MCIs?

It prioritizes patients based on injury severity to maximize survival rates.

What system is used for pediatric triage?

JumpSTART.

How do triage tags assist in patient management?

They provide visual categorization of patients' conditions and treatment priorities.

What personal protective measures should responders take at hazardous MCI scenes?

Wear appropriate PPE and conduct environmental assessments to mitigate risks.

Chapter 9: Fireground Rescues: Overcoming Smoke, Heat, and Hazardous Conditions

Performing fireground rescues demands a deep understanding of how to navigate the perilous conditions that arise during a fire. Rescuers are often tasked with working amidst intense heat, thick smoke, and structurally compromised environments, each presenting unique hazards. The ability to effectively maneuver through these conditions is crucial not only for the successful retrieval of individuals in danger but also for ensuring the safety and survival of the rescuers themselves. It requires a combination of physical skill, strategic thinking, and thorough preparedness, enabling firefighters to adapt swiftly to the dynamic nature of firefighting scenarios. This chapter delves into essential techniques and approaches that create a foundation for mastering the art of fireground rescue while maintaining high safety standards.

Within the scope of this chapter, several critical aspects of fireground rescues will be thoroughly explored. Readers will gain insights into distinguishing fire suppression tasks from dedicated rescue operations, both of which necessitate precise coordination yet have distinct objectives. Additionally, the chapter highlights the importance of recognizing different building types, such as residential, commercial, and industrial, which present varying levels of complexity and risk. The discussion extends to overcoming extreme conditions like limited visibility and structural instability, which require strong situational awareness and the use of technological aids. Effective communication strategies, pivotal in overcoming noise and rapidly changing conditions, will also be examined, alongside methods for risk assessment and decision-making. Finally, the chapter considers advanced search and rescue techniques designed for diverse scenarios, including high-rise buildings, equipping firefighters and first responders with the knowledge needed to undertake their roles efficiently and safely.

Introduction to Fireground Rescues

When it comes to fireground rescues, understanding the foundational concepts is critical for ensuring the safety of both victims and first responders. The distinction between fire suppression tasks and dedicated rescue operations is a primary consideration. While fire suppression focuses on controlling and extinguishing fires, rescue operations prioritize locating and removing individuals from danger. Both require strategic coordination but serve different objectives.

Fire suppression involves deploying hoses, assessing water supply, and managing elements that can exacerbate a fire, such as fuel sources and oxygen. Rescue operations, on the other hand, demand rapid assessment and movement within hazardous environments to find and evacuate victims. This dual tasking often requires firefighters to switch roles seamlessly, making an in-depth understanding of each role essential (A Firefighter's Guide to Fireground Search and Rescue – Part 1, n.d.).

The types of rescues encountered vary by the nature of structures involved, including residential, commercial, and industrial buildings. Residential fires are common and might involve navigating complex layouts with numerous rooms and personal items that could hinder accessibility. Commercial buildings add layers of complexity due to larger spaces, heavier materials, and specialized equipment or chemicals present in some businesses. Industrial facilities present even more heightened risks such as hazardous materials and massive structural features that could collapse unexpectedly.

Each scenario demands tailored strategies. For example, residential rescues often focus on bedrooms or living areas where people are most likely to be found. Commercial rescues might target offices or assembly areas, while industrial settings may require dealing with machinery or hazardous substances. Recognizing these distinctions enables responders to adapt their approach swiftly and effectively (StackPath, n.d.).

A significant challenge in fireground rescues lies in overcoming extreme conditions like intense heat, limited visibility, and unstable structures. Firefighters frequently contend with temperatures that can reach hundreds of degrees, necessitating protective gear and cooling techniques. Visibility issues, often compounded by thick smoke, make it difficult to locate both victims and potential exit routes. Such conditions demand strong situational awareness and reliance on tools like thermal imaging cameras.

Structural instability poses grave concerns during rescues. Fires can weaken building supports, leading to potential collapses. Firefighters must conduct thorough assessments, using pre-planned knowledge of building layouts when available, to navigate these dangers safely. Understanding the signs of impending structural failure, such as creaking sounds or sagging ceilings, is crucial for avoiding unnecessary risks (A Firefighter's Guide to Fireground Search and Rescue – Part 1, n.d.).

Communication difficulties further complicate rescue efforts. Equipment noise, including the hum of breathing apparatus and the crackle of radios, can obscure vital information. Furthermore, dynamic fire conditions mean that situations can shift rapidly, demanding constant updates. Effective communication protocols are necessary to ensure that teams stay informed and aligned in their objectives. Clear command channels and redundant systems help mitigate misunderstandings and delays that could endanger lives.

Adapting communication methods to overcome these barriers is essential. Utilizing hand signals, pre-established codes, and repeated message protocols can enhance clarity. Additionally, integrating advanced technologies like wireless transmitters or earpieces into firefighting gear may improve the ability to relay messages within noisy environments (StackPath, n.d.).

Fire Behavior and Its Impact on Rescues

Fire behavior is a critical element that influences the success and safety of fireground rescues. As firefighters, recognizing how certain dynamics like fuel availability, heat, and oxygen interaction—collectively known as the fire triangle—affect rescue operations is essential for effective firefighting. The fire triangle serves as a foundation for understanding fire development phases and guides firefighters in anticipating changes during rescue missions. Fuel provides the necessary material for combustion, and its availability often dictates the intensity and spread rate of a fire. Heat catalyzes the combustion process, creating energy that further fuels fire growth, while oxygen supports the chemical reaction necessary for sustaining flames.

As fires progress through their developmental stages, understanding these interactions can aid firefighters in making informed decisions. For instance, in the incipient phase, the fire is typically small and contained, offering better conditions for rapid intervention. In contrast, the fully developed stage presents greater challenges due to rampant heat and smoke, necessitating strategic adjustments in rescue tactics (Extreme Fire Behavior: Understanding the Hazard | CTIF - International Association of Fire Services for Safer Citizens through Skilled Firefighters, n.d.).

Moreover, acknowledging flashover and backdraft risks is vital for firefighter safety and rescue effectiveness. Flashover represents a critical transition point where almost all combustible materials in an area ignite simultaneously due to intense heat. Key indicators include high temperatures, darkening windows, and rollover phenomena, signaling the need for immediate evacuation or defensive strategies to prevent firefighter entrapment (Unlocking the Secrets of Fire Behavior: A Vital

Evolution in Firefighting Tactics, 2024). Meanwhile, backdraft poses an equally lethal threat, occurring when a fire is starved of oxygen and subsequently reignites explosively upon air re-introduction. Warning signs such as yellowish smoke, pulsing doors or windows, and low rumbling sounds often precede backdraft episodes, underscoring the necessity for vigilant assessment before entry.

Tactical ventilation plays a pivotal role in altering fire behavior and enhancing rescue operations. By systematically controlling airflow within a structure, firefighters can influence fire progression and create safer environments for occupant evacuation. Properly executed ventilation helps stabilize interior heat levels, reduces smoke accumulation, and improves visibility, allowing for more efficient victim search and extrication. However, improper or untimely ventilation can exacerbate fire growth, emphasizing the importance of situational awareness and coordination in executing these tactics effectively (Extreme Fire Behavior: Understanding the Hazard | CTIF - International Association of Fire Services for Safer Citizens through Skilled Firefighters, n.d.).

The impact of water application on smoke conditions and structural integrity further highlights the complex relationship between suppression techniques and rescue outcomes. Strategically applied water can dramatically alter smoke density and toxicity levels, enabling safer navigation through compromised environments. Effective water use also mitigates thermal effects, reducing the likelihood of flashover scenarios and supporting overall structural stability. Nevertheless, excessive water usage risks structural compromise due to added weight and potential integrity loss in weakened materials, warranting careful assessment by firefighting teams.

Fireground Size-Up and Risk Assessment

Conducting effective size-up and risk assessment is vital for successful fireground rescues. The ability to rapidly assess a scene ensures that firefighters can make informed decisions that protect both themselves and those they're rescuing.

The first step in this process is performing an initial scene evaluation. As the first responders arrive at the scene, they should immediately determine the status of occupants. Are there signs of life or distress signals? This initial assessment helps prioritize actions. Equally important is assessing the fire spread. Identifying where the fire is most intense and how quickly it appears to be moving allows for strategic planning of entry points and rescue routes. Ensuring accurate information feeds into the broader tactical decisions that will follow.

Understanding smoke conditions is another critical skill for firefighters. Smoke analysis provides clues about the fire's location and dynamics. For instance, thick black smoke typically indicates a hotter, more advanced fire stage, often involving synthetic materials or heavy fuels (Staff, 2022). In contrast, white or light grey smoke might suggest that the fire is in its early stages or that less dense materials are burning. Observing the velocity and density of the smoke also reveals the pressure inside the building, informing firefighters about potential temperature escalations. These observations can prevent situations where the team inadvertently enters areas with backdraft potential—an explosive event all firefighters aim to avoid.

Next, evaluating structural stability cannot be overstated. Fire weakens structural elements, causing potential collapse hazards. Buildings with compromised integrity pose significant risks during rescues. Firefighters must recognize warning signs such as sagging floors, warped door frames, or unusual noises like creaking or popping. Regular training on different building types is essential since responses differ based on construction materials and design. Residential buildings behave differently under stress compared to commercial structures. With preplanning and real-time assessments, crews can anticipate dangers and adjust their operations accordingly.

Prioritizing rescues by weighing firefighter safety against mission objectives represents a fundamental guideline for decision-making. Rescue prioritization involves assessing which actions can be safely executed in alignment with the mission goals. Guidelines here become crucial, particularly in high-risk environments (Staff, 2018). Decisions need to be made swiftly, often balancing between saving lives and minimizing risk to the rescuers. This may sometimes require discerning whether direct interior attacks are feasible or if defensive tactics should be employed until conditions improve.

In scenarios where there are multiple victims or complex rescue challenges, this prioritization becomes even more important. Tactical decision-making tools can aid officers in determining the sequence of rescues. Guidelines help teams focus on savable lives while continuously reassessing the situation. This reassessment includes employing models like the Value, Time, Size (VTS) model, which is a structured approach to decide whether an offensive or defensive strategy should be deployed. The VTS model considers the value of potential lives saved, the available time to act, and the size or extent of resources needed (Staff, 2018). If any aspect indicates excessive risk without proportional benefit, it suggests a shift to defensive measures.

Fire scenes are dynamic and require adaptable strategies. Teams should foster open communication, ensuring that updates on fire behavior, structural changes, and occupant discoveries are shared instantly among all firefighters involved. Advanced tools, such as thermal imaging cameras and personal protective equipment, although not directly discussed in detail here, support these assessments and reinforce the safety net for rescue operations.

Search and Rescue Techniques in Fire Conditions

In fireground rescues, mastering search and rescue techniques is critical for ensuring the safety of both victims and firefighters. Amidst smoke, heat, and hazardous conditions, understanding the distinction between primary and secondary searches is fundamental to effective operations.

A primary search is a swift, systematic approach that occurs during or immediately after fire suppression efforts to locate victims who might be in immediate danger. These searches focus on areas most likely impacted by fire, such as rooms adjacent to the fire source or floors directly above it, thereby enhancing the chances of rescue success. Primary searches rely on recognized patterns like wall contact and directional turns, carried out by two-person teams maintaining visual, voice, or physical contact ((<i>A Firefighter's Guide to Fireground Search and Rescue – Part 2</i>, n.d.)).

Once immediate threats are addressed, secondary searches commence. Unlike their primary counterparts, secondary searches occur when the situation is controlled, allowing rescuers to meticulously comb through debris and explore any overlooked areas for victims. The methodical nature of secondary searches ensures all spaces, including hidden enclaves, undergo thorough examination before marking them as clear ((<i>Firefighter Search and Rescue Manual & Tips 2021</i>, n.d.)).

Implementing efficient search patterns enhances the speed and effectiveness of rescue missions. Techniques such as the oriented search method provide a structured approach within constrained environments. Here, one firefighter remains anchored with visibility aids, while others explore remaining areas, maintaining consistent procedures by checking door temperatures, anticipating fire spread, and devising exit strategies. Another essential technique is Vent-Enter-Isolate-Search (VEIS), designed for rapid appraisal of high-risk zones, like bedrooms above fires. In VEIS, targeted venting followed by isolation promotes focused searches and prompts swift victim retrieval

(<i>A Firefighter's Guide to Fireground Search and Rescue – Part 2</i>, n.d.) ([Learn more: How to conduct firefighter primary search training]).

Thermal imaging technology represents an invaluable asset, augmenting traditional search methodologies. Thermal imagers offer clarity amidst dense smoke and enable precise identification of potential victims by highlighting their heat signatures. Additionally, these devices assist in assessing fire behavior, granting commanders real-time data to adapt strategies efficiently and potentially averting catastrophic scenarios ((<i>Firefighter Search and Rescue Manual & Tips 2021</i>, n.d.)).

Victim removal methods require finesse and situational awareness. Often, unconscious individuals necessitate specialized extraction techniques, such as drags and carries. Dragging involves positioning the victim's arms over the rescuer's shoulders or underarms, providing stability during evacuation. Carries, on the other hand, may involve the cradle carry for children or the firefighter lift for adults. Regardless of method, ensuring the airway remains open and limiting movement where possible reduce the risk of further injuries during removal ((<i>A Firefighter's Guide to Fireground Search and Rescue – Part 2</i>, n.d.)).

Effective execution of search and rescue operations hinges on adherence to predefined guidelines, communication, and teamwork. For instance, maintaining constant communication with command posts ensures rescuers receive timely updates on changing conditions, which is vital for adapting search strategies. Moreover, working collaboratively, often in pairs or small groups, fosters accountability and safety, reducing risk and increasing operational efficacy ((<i>Firefighter Search and Rescue Manual & Tips 2021</i>, n.d.)).

While mastery of specific techniques is crucial, a comprehensive understanding of the dynamic nature of firegrounds empowers rescuers to make informed decisions. Scenarios evolve rapidly; thus, continuous training and familiarization with structures enhance adaptability and proficiency. Drills simulating diverse fire scenarios develop reflexive skills, enabling teams to innovate within standard procedures when confronting atypical challenges ((<i>A Firefighter's Guide to Fireground Search and Rescue – Part 2</i>, n.d.)).

High-Rise and Large-Structure Fire Rescues

High-rise and large-structure fire rescues present distinct challenges that demand strategic planning and execution. These environments often mean extended response times due to their sheer size and complexity, which can hamper immediate access for firefighters. Buildings above 120 feet, categorized as high-rises, present substantial egress difficulties (Aumond, 2024). In these scenarios, the failure of fire safety systems can result in compromised evacuation efforts, as tragically demonstrated by incidents like the Grenfell Tower fire (Aumond, 2024).

One critical aspect of overcoming these challenges is the use of stairwells for fire attack and rescue operations. Stairwells serve as primary routes for both evacuating occupants and advancing hose lines safely during a rescue operation. Ensuring these stairways are strategically located and properly sized is crucial, reducing the risk of congestion during evacuation while providing clear pathways for emergency responders (Graham, 2023). Additionally, these areas must be fortified with pressurization systems and fire-rated barriers to prevent smoke infiltration, ensuring a safe and controllable environment for movement (Aumond, 2024). Signage plays a vital role too; prominent exit signs and comprehensive egress plans must be in place, guiding occupants reliably toward exits during an emergency.

In terms of coordinating fire attacks and rescues, the interaction between firefighters and built-in safety features such as fire service access elevators becomes pivotal. These elevators allow first responders rapid access to floors higher up in the building, significantly improving their operational efficiency (Aumond, 2024). Strategically using these elevators enables first responders to set up hoses faster and direct firefighting efforts more effectively. Furthermore, these elevators, integrated with interior exit stairways, offer reliable egress options for both rescuers and occupants when needed.

Decision-making in high-rise fires often involves weighing the options between shelter-in-place and full evacuation strategies. This decision hinges on several factors, including the specific dynamics of the building and the behavior of the fire itself. The phased evacuation strategy reminds us that sometimes sheltering in place may avoid unnecessary exposure to hazards (Aumond, 2024). An emergency voice alarm communication (EVAC) system is instrumental here, relaying clear instructions and coordinating phased evacuations by targeting specific zones within the building (Graham, 2023).

Establishing safe refuge zones is another crucial approach to mitigating mass casualties during high-rise fires. Refuge areas provide temporary safety for building occupants who cannot immediately evacuate or when conditions outside become untenable. Such zones should be equipped with enhanced ventilation and pressurization systems to maintain breathable conditions free of hazardous smoke (Aumond, 2024). Planning these spaces requires careful consideration of their location within the building, ideally close to core structural elements that provide additional protection.

Finally, clear communication and coordination among all parties involved are fundamental to the success of high-rise rescue operations. Implementing robust BDA systems enhances radio signal strength, facilitating seamless communication between first responders even in challenging high-rise environments (Aumond, 2024). Regular training and realistic fire drills further prepare occupants, enhancing their understanding of evacuation protocols and empowering them to respond promptly when alarms sound (Graham, 2023).

Summary and Reflections

In this chapter, we explored the essential techniques and challenges involved in performing effective fireground rescues. Emphasizing the critical distinction between fire suppression and rescue operations, we identified the need for firefighters to rapidly adapt between roles of controlling fires and evacuating individuals under extreme conditions. We analyzed various rescue scenarios within residential, commercial, and industrial settings, each requiring tailored strategies to navigate complex layouts and address specific risks such as hazardous materials and structural instability. Understanding fire behavior, including dynamics like fuel availability and oxygen interaction, is paramount for anticipating changes during rescue missions.

Moreover, essential considerations such as communication and risk assessment were addressed for conducting successful rescues. We discussed overcoming barriers, like equipment noise and unstable structures, through effective protocols and technologies, such as thermal imaging cameras. The importance of size-up and situational awareness was emphasized for ensuring firefighter safety while prioritizing rescues based on mission objectives. Tactics for high-rise and large-structure firefighting were examined, highlighting strategic use of stairwells, elevators, and communication systems. Through mastering these comprehensive skills, firefighters can enhance their proficiency, making informed decisions that prioritize the safety of both victims and their team.

Chapter 9 Test Questions and Answers

What is the primary difference between fire suppression and rescue operations?

Fire suppression focuses on controlling and extinguishing fires, while rescue operations prioritize locating and removing individuals from danger.

What are the key challenges of fireground rescues?

Fireground rescues are complicated by extreme conditions such as intense heat, limited visibility, unstable structures, and communication difficulties.

Why is fire behavior crucial in rescue operations?

Understanding fire behavior, including the fire triangle (fuel, heat, and oxygen), helps firefighters anticipate changes and adjust their tactics to improve safety during rescues.

What factors differentiate residential, commercial, and industrial rescues?

Residential rescues often involve navigating complex layouts; commercial rescues deal with larger spaces and specialized equipment, while industrial rescues may involve hazardous materials and massive structural features.

What are the risks posed by flashover and backdraft in fire rescues?

Flashover occurs when all combustible materials ignite simultaneously due to intense heat, and backdraft happens when a fire is starved of oxygen and then explodes upon reintroduction of air.

What role does tactical ventilation play in fireground rescues?

Tactical ventilation can control fire behavior by reducing heat, smoke, and improving visibility, which allows for safer environments for both rescues and firefighting efforts.

How can water application impact fireground rescues?

Proper water application helps reduce smoke density and temperature, making navigation easier, but excessive use may compromise structural integrity.

What is a primary search in fireground rescues?

A primary search is a swift, systematic search to locate victims in immediate danger, often conducted during or immediately after fire suppression efforts.

What differentiates a secondary search from a primary search?

A secondary search occurs once the situation is controlled, focusing on thoroughly combing through debris and hidden spaces to ensure no victims are missed.

What is the Vent-Enter-Isolate-Search (VEIS) technique?

VEIS is a rapid search method where firefighters vent an area, enter it, isolate the fire, and then search for victims, particularly in high-risk zones like bedrooms above fires.

Why is thermal imaging technology important in fireground rescues?

Thermal imaging allows firefighters to see through smoke, helping identify victims by detecting heat signatures and assisting in fire behavior assessment.

What are common methods for victim removal during rescues?

Victim removal often involves dragging or carrying techniques, such as the firefighter lift or cradle carry, depending on the victim's condition.

How do structural instability and fire conditions affect rescues?

Fires can weaken building supports, leading to potential collapses. Firefighters must be aware of warning signs of structural failure, such as sagging ceilings or unusual noises.

What is the importance of conducting an initial scene evaluation?

An initial scene evaluation helps assess the status of occupants, fire spread, and smoke conditions, which are crucial for prioritizing actions and planning rescue strategies.

What role does smoke analysis play in fireground rescues?

Smoke analysis provides clues about fire intensity and location. For example, thick black smoke often indicates a more advanced fire, while white smoke suggests an early stage.

How should firefighters prioritize rescues during fireground operations?

Rescues should be prioritized by assessing the risks to firefighters and victims, determining which actions can be safely executed, and continuously reassessing the situation.

What is the Value, Time, Size (VTS) model, and how is it used?

The VTS model helps firefighters decide between offensive or defensive strategies by considering the value of lives saved, available time, and resources needed for a successful rescue.

What challenges do high-rise and large-structure fire rescues present?

These buildings present challenges such as extended response times, difficulties in access, and the need for specialized rescue systems like fire service access elevators and stairwell management.

What are safe refuge zones, and why are they critical in high-rise fires?

Safe refuge zones are areas within buildings that provide temporary safety for occupants who cannot evacuate immediately. They should be equipped with enhanced ventilation and pressurization systems.

How does communication impact the success of high-rise fire rescues?

Clear communication is essential in high-rise rescues, with tools like BDA systems improving radio signal strength and ensuring seamless coordination among first responders during the operation.

Chapter 10: Training for Excellence: Drills, Simulations, and Continuous Improvement

Training for excellence in rescue operations is a comprehensive endeavor that combines drills, simulations, and continuous improvement to build the utmost proficiency among first responders. The process starts with thorough preparation, focusing on equipping rescue teams with essential skills to act swiftly during emergencies. This chapter delves into how strategic training methods enable firefighters and other first responders to make quick, informed decisions when it matters most. By embedding these skills into routine practice, teams become adept at minimizing response times and maximizing their impact during critical situations.

The chapter intricately explores various components of effective training programs, highlighting the significance of preparedness through practical exercises. Readers will discover how targeted decision-making training refines cognitive abilities, ensuring fast assessment and resolution of complex scenarios. The role of muscle memory developed through repetitive drills is examined, demonstrating its importance in executing tasks efficiently under pressure. Additionally, this section addresses the psychological resilience fostered by realistic simulations, enabling responders to manage stress and maintain composure amid chaos. By integrating these elements, the chapter offers a roadmap for cultivating an environment of continuous learning and adaptation, essential for evolving rescue demands.

Introduction to Training in Rescue Operations

Training plays a pivotal role in the efficiency and success of rescue operations. By equipping teams with the necessary skills, knowledge, and readiness, training can significantly reduce response times. A well-prepared team is capable of acting swiftly, thus enhancing their ability to make a positive impact during emergency situations. This preparedness does not happen overnight; it is a result of continuous learning and practice.

One of the key benefits of regular training is the enhancement of preparedness and efficiency. Response time is crucial in rescue operations, often being the difference between life and death. Effective training ensures that rescuers are familiar with the necessary protocols and procedures, which enables them to execute their duties promptly. Teams that are well-trained can minimize delays by immediately identifying their roles, understanding their objectives, and efficiently implementing the steps required to resolve an emergency. Such precision comes from repeated exposure to various scenarios, allowing responders to act instinctively when under pressure. Creating realistic scenarios familiarizes trainees with the exact dynamics they might face, which builds confidence and assures quick, decisive actions.

In addition to swift response, coordination improves dramatically under pressure through targeted decision-making training. Effective communication and collaboration among team members can be significantly enhanced when individuals are trained to prioritize tasks and share information succinctly under stressful conditions. Training programs should incorporate simulations and role-playing exercises that replicate these high-pressure environments. As Calleja highlighted in Source 2, using interconnected simulators fosters teamwork and strategic thinking, which is essential for real-life mission success (Training for Any Emergency – without the Risk, 2025). This type of training enables each team member to understand their unique role within the broader operation, reducing miscommunication and confusion.

Targeted decision-making training goes beyond just understanding individual responsibilities. It encourages critical thinking and enables teams to evaluate situations quickly, consider possible outcomes, and decide on the best course of action. This approach minimizes hesitation and optimizes the use of resources in urgent situations. Decision-making under duress is enhanced through scenario-based training, where trainees learn to analyze problems and implement solutions rapidly. These exercises ensure that decision-making processes become second nature, improving reaction times and effectiveness during actual emergencies.

Another significant factor contributing to successful rescue operations is the application of muscle memory through repetition under stress. The principle behind this approach is simple: frequent practice of specific actions engrains them into the subconscious, making them automatic responses during emergencies. For instance, repeatedly practicing donning protective gear or setting up equipment reduces the time taken during real events, as actions are executed without conscious thought. Muscle memory ensures that even in high-stress situations, rescuers can perform their duties efficiently and accurately.

Repetitive training also aids in reinforcing the correct techniques and procedures, minimizing errors and improving overall proficiency. By embedding these actions into muscle memory, rescuers are less likely to falter under pressure, ensuring that their performance remains consistent regardless of the situation's intensity. Incorporating stress elements into training sessions, such as timed drills or unexpected challenges, further enhances this approach's effectiveness. Trainees become accustomed to operating under stress, thereby developing resilience and maintaining calmness, even in chaotic environments.

Realistic training scenarios are instrumental in providing psychological benefits that enhance performance. Simulations that closely mimic the conditions of potential emergencies expose trainees to the mental pressures associated with real-life rescue missions. As noted in Source 1, realistic simulations immerse participants in lifelike situations, helping them understand their roles' gravity and responsibilities (RNR Rescue, 2023). This experience helps build mental toughness and prepares individuals to handle the emotional challenges they may encounter.

Beyond mental toughness, realistic scenarios develop adaptability and creativity, crucial traits in dynamic rescue operations. Emergencies rarely unfold according to plan, so trainees must learn to think on their feet and devise innovative solutions to unforeseen obstacles. By facing a variety of simulated challenges, rescuers hone these skills, becoming more adept at adjusting their strategies and maintaining control in unpredictable situations.

Moreover, realistic scenarios help inoculate teams against stress. Repeated exposure to adverse conditions during training accustoms them to pressure, reducing the likelihood of panic or paralysis during actual operations. This stress inoculation effect boosts confidence, equipping team members with the self-assurance needed to tackle any crisis effectively. As Source 1 indicates, the confidence gained from these exercises translates into better performance in real emergencies because it convinces rescuers of their capability to manage tough situations proficiently (RNR Rescue, 2023).

Core Competencies in Rescue Training

In the realm of rescue operations, possessing specific skills is essential to ensure successful outcomes when lives are at stake. These skills can be categorized into four pivotal areas: physical conditioning, rapid risk assessment and critical thinking, mastery of technical tools, and effective communication techniques. Each area contributes significantly to the overall efficiency and safety of rescue missions.

First and foremost, physical conditioning is crucial for rescuers who often face demanding situations that require both strength and stamina. Rescue missions can involve traversing rugged terrains, climbing buildings, or carrying heavy equipment and individuals. Therefore, maintaining peak physical fitness is not just beneficial but a necessity. Regular exercise regimens focusing on cardiovascular endurance, muscular strength, and flexibility can prepare team members for these physically taxing scenarios. According to Piatt (2023), consistent training activities help build the confidence and competence needed in real-world emergencies, ultimately reducing response times.

Moving on to the cognitive demands of rescue operations, rapid risk assessment and critical thinking are invaluable skills every rescuer must develop. Situational awareness is key in any rescue scenario; being able to quickly evaluate risks, anticipate potential hazards, and make informed decisions can save lives. Education in search and rescue fosters these critical thinking abilities, allowing responders to adapt to ever-changing environments and solve complex problems efficiently (Education - National Association for Search and Rescue, n.d.). In high-pressure settings, where every second counts, the ability to think clearly and act decisively can be the difference between success and failure.

Equally important is the mastery of technical tools that modern rescue operations rely upon. Equipment such as self-contained breathing apparatuses (SCBA) and thermal imaging cameras are integral to navigating and managing various emergency situations. These tools enable rescuers to access hazardous environments safely and locate individuals in need of assistance more effectively. Proficiency with these devices ensures that teams can operate efficiently under challenging conditions and maximize their chances of successful outcomes. As Piatt (2023) suggests, understanding and mastering rope rescue techniques and other specialized equipment are foundational elements of comprehensive training programs.

Another cornerstone skill set in rescue operations is communication. Effective communication techniques are vital for fostering teamwork and ensuring smooth coordination during stressful situations. Miscommunication can lead to confusion, delays, and even jeopardize mission success. Establishing clear protocols, roles, and responsibilities enhances the team's ability to function as a cohesive unit. Implementing Search and Rescue (SAR)-specific software and tools like GPS trackers and incident command systems further improves coordination, enabling rescuers to communicate seamlessly even in remote or severe weather conditions (Piatt, 2023). Moreover, regular drills focused on communication allow teams to practice and solidify these protocols, ensuring readiness when it matters most.

While mastering these skills is vital, balancing them with the operational demands faced by rescue teams is equally critical. Physical conditioning and endurance require continuous effort and time. Creating structured workout plans that integrate into daily routines can help rescuers maintain their fitness levels without compromising their operational responsibilities. Similarly, ongoing education and training in risk assessment, problem-solving, and technical proficiency should fit around real-world mission schedules. This integration ensures that when the call for action arises, team members are prepared and equipped with the necessary skills and knowledge.

Structuring a Comprehensive Training Program

Guiding the creation of structured and effective training schedules in rescue operations begins with aligning all plans—whether annual, monthly, or weekly—with real-world needs. At the core of these training schedules should be a focus on the dynamic nature of real-life emergencies, ensuring that the skills developed are relevant and applicable. For instance, an annual plan might involve comprehensive reviews and assessments to ensure alignment with new tools, technologies, and procedures that have been introduced over the course of the year. Monthly schedules can then break down these larger objectives into more manageable modules or workshops, allowing for focused learning of specific skills.

Weekly plans are essential for maintaining momentum and allowing for continuous improvement. They should involve hands-on practice through drills and simulations that replicate actual rescue scenarios as closely as possible. This regular engagement not only helps in reinforcing lessons learned but also ensures that firefighters and first responders are always prepared for emergencies by staying current with their training. Incorporating exercises that simulate real-world conditions enhances readiness and adaptability, which are critical in life-or-death situations.

Evaluation is another crucial component that ensures training remains effective and beneficial. Regularly evaluating progress allows training programs to adapt to changing needs or unexpected challenges. Performance benchmarks serve as important indicators of whether training objectives are being met. These benchmarks should include specific measures related to speed, accuracy, safety, and effectiveness in various simulated rescue situations. By setting clear standards, these benchmarks drive consistent improvement and highlight areas where additional focus may be needed.

Implementing a system for curriculum adjustments based on these evaluations is vital. Adjustments may involve updating the content to incorporate new findings from ongoing research, feedback from previous evaluations, or changes in organizational priorities. This process makes certain that training material stays relevant and continues to meet the demands of the ever-evolving landscape of rescue operations.

Competency levels must be assessed using practical and scenario-based tests to gauge the effectiveness of training. These tests should replicate the complexities and pressures of real emergency situations to truly assess a participant's readiness. Practical assessments could involve navigating typical rescue mission obstacles, utilizing equipment under challenging conditions, or recognizing and responding to unexpected developments during a simulation. Scenario-based testing not only evaluates the technical ability but also how well individuals can apply critical thinking to solve problems under stress.

To ensure these competencies translate effectively into the field, organizations could adopt the Kirkpatrick Model as a framework for evaluation. Through systematic assessments at different stages, such as participant reactions, learning outcomes, behavior changes, and results, it becomes possible to gain valuable insights into training efficacy (SOLUTION, 2024).

Moreover, integrating feedback loops allows for ongoing enhancements to the training processes. This involves regularly collecting data from trainees' experiences and outcomes from scenarios, and then feeding this data back into the planning and execution phases. This ensures a cycle of continuous improvement, with every iteration of the training program building on past successes and correcting identified shortcomings.

An illustrative example of successful application of benchmarks and assessments is seen in customer service training programs where key performance indicators (KPIs), like customer satisfaction scores, are used to measure impact. Such an approach in rescue operations could similarly help quantify training success and pinpoint specific areas for development. Over time, comparisons of pre- and post-training assessments will visibly show growth in knowledge and efficiency, providing concrete evidence of the program's effectiveness (Explorance, 2023).

In crafting robust training schedules, it's important to plan strategically to address both individual and team development. While individual competencies are critical, team effectiveness hinges on how well members coordinate and communicate during an operation. Including elements that enhance situational awareness and decision-making skills within the training can greatly benefit team dynamics during real rescues. Practical guidelines can support achieving these objectives; for instance, incorporating regular debrief sessions after training exercises facilitates reflection and collective problem-solving among team members.

Furthermore, physical conditioning elements focusing on strength and stamina are integral to preparing participants for the physically demanding nature of rescue missions. Exercises aimed at enhancing flexibility and preventing injuries should be part of the training regime, promoting long-term health and capability among rescuers.

Live Drills and Simulations

The effectiveness of live drills and simulations in rescue operations lies at the core of equipping first responders with essential skills and readiness to face real emergencies. In this section, we will explore how these training methods foster preparedness and operational efficiency.

Live drills are a cornerstone of emergency preparedness training, encompassing various scenarios such as structural firefighting and vehicle extrication. These exercises focus on developing practical skills that ensure readiness for different types of emergencies. Structural firefighting drills, for instance, enable firefighters to practice navigation through smoke-filled environments, reinforcing their ability to manage challenging conditions safely and efficiently. Similarly, vehicle extrication drills provide crucial experience in using specialized tools to free trapped victims promptly, enhancing the response capability during road accidents or similar incidents.

To maximize the effectiveness of these drills, incorporating realistic elements plays a significant role. By simulating true-to-life conditions, such as intense fire scenarios or scattered debris, the training process becomes more immersive and practical. This exposure helps participants adapt to the unpredictable nature of actual emergencies, fostering better situational awareness and critical decision-making skills. For example, introducing controlled fires and lifesaving mannequins can offer a tactile understanding of heat dynamics and victim handling techniques in realistic settings.

While realism is vital, safety protocols must remain paramount during these drills. The presence of safety officers ensures strict adherence to safety guidelines, minimizing risk and preventing injuries. Their role involves overseeing the drill's conduct, monitoring environmental conditions, and providing immediate intervention when necessary. By maintaining a controlled and secure environment, trainees can focus on experiencing the challenges without compromising their safety. Deploying safety officers also facilitates immediate feedback and post-drill evaluations, allowing individuals to reflect on performance and areas for improvement.

In conjunction with live drills, simulation-based training offers an innovative approach to preparing first responders without exposure to danger. Virtual Reality (VR) technology, increasingly utilized in emergency training, provides a resourceful medium for honing decision-making and problem-solving skills in risk-free environments. VR simulations enable users to engage in dynamic scenarios, from managing hazardous material spills to executing building evacuations. This form of training supports repetitive practice, allowing participants to refine their strategies and responses across varied scenarios without resource constraints common in physical drills.

Furthermore, simulation exercises allow trainers to customize scenarios to address specific training needs or current threats effectively. For instance, with VR, users can practice coordination during multi-agency responses or test communication systems' robustness under stress. Customizable scenarios also offer flexibility in updating training modules to incorporate new methodologies or integrate lessons learned from recent events.

Practical training, whether through live drills or simulations, requires a balance between technical proficiency and teamwork. Mastery of tools such as extrication devices and Self-Contained Breathing Apparatus (SCBA) is crucial, ensuring individuals perform tasks safely and effectively. These skills are paramount, not just for successful task execution but also for maintaining personal safety in volatile situations. Adequate technical training prepares responders to utilize equipment optimally while navigating complex scenarios.

Equally important is nurturing teamwork and communication under pressure. Training scenarios replicate high-stress environments where efficient verbal and non-verbal communication techniques significantly impact outcomes. Practicing communication ensures clarity and coordination among team members, reducing response times and enhancing operational cohesion. As seen in both live and simulated drills, fostering these skills builds trust within teams and promotes a culture of collaborative problem-solving.

Continuous improvement remains a fundamental objective, demanding regular evaluation and feedback integration into training regimens. Post-drill debriefings offer invaluable insights, facilitating open discussions on what went well and identifying areas needing enhancement. Encouraging participant feedback cultivates a learning-oriented atmosphere, driving ongoing refinement of practices and readiness levels. Evaluating drills comprehensively leads to informed adjustments in emergency protocols, advancing organization-wide emergency preparedness efforts.

Cross-Disciplinary and Specialized Training Programs

In today's rapidly evolving landscape of emergency response, the importance of advanced and collaborative training methodologies cannot be overstated. These techniques not only foster teamwork across different agencies but also ensure that personnel are equipped with specialized skills to handle diverse rescue scenarios. By diving into these methodologies, responders can achieve greater efficiency and effectiveness in times of crises.

1. Enhancing Interoperability through Coordinated Training

Interoperability between various emergency services such as fire departments, emergency medical services (EMS), and police is crucial for seamless operation during multi-agency responses. Collaborative training sessions are essential to enhance this interoperability by bringing teams together to practice joint operations. When these agencies work side-by-side in controlled training environments, it leads to a shared understanding of each unit's capabilities and constraints. This preparation ensures that during actual emergencies, communication flows smoothly, roles are clear, and tasks are executed efficiently.

Guideline: During coordinated training, integrate representatives from fire, EMS, police, and other agencies to simulate real-world operations and create protocols that are practiced regularly. To maintain optimal readiness, it's crucial to develop training plans that are structured on annual, monthly, and weekly bases, ensuring continuous skill development across all participating emergency services.

2. Specialized Programs for Technical Skill Development

Rescue operations often demand specific technical skills beyond fundamental emergency response abilities. One such area is rope rescue, essential in scenarios where victims are trapped in elevated or difficult-to-reach locations. Specialized programs focusing on technical skills provide targeted training to address these demands. Personnel learn to use equipment like harnesses, pulleys, and climbing ropes expertly, allowing them to execute complex rescues safely and efficiently.

These programs often involve practical hands-on sessions coupled with theoretical knowledge, providing a comprehensive learning experience. By focusing on such technical skill development, responders not only broaden their expertise but also boost their confidence in handling high-stake situations.

3. Integrating Urban and Wilderness Rescue Methods

A notable aspect of training advancement is the integration of urban and wilderness rescue methods. Joint exercises involving Search and Rescue (SAR) teams bridge the gap between these two distinct environments. Urban rescues often focus on navigating through debris or collapsed structures, while wilderness rescues might require navigation over rugged terrains or dense foliage.

Through combined training exercises, responders gain versatility, allowing them to adapt strategies suited for varied landscapes. These joint exercises teach participants how to apply various techniques, whether they are rescuing someone from a flooded city street or extracting a hiker stranded on a mountain trail. Knowledge-sharing among teams during these exercises enriches everyone's toolkit, ensuring preparedness regardless of the operational environment.

4. Preparing for High-Risk Situations: Hazmat and Active Shooter Drills

Another critical component of modern training methodologies is preparing responders for high-risk situations like hazardous material (hazmat) incidents and active shooter scenarios. Drills dedicated to these events equip personnel with the necessary skills to act swiftly and safely while minimizing risk to themselves and the public.

Hazmat training involves familiarizing responders with protective gear, decontamination procedures, and chemical identification tools. It emphasizes quick decision-making, efficient evacuation processes, and containment strategies. Similarly, active shooter drills prepare teams for tactical responses, prioritizing the neutralization of threats while safeguarding civilians.

Conducting these drills regularly ensures that responders remain attuned to the latest developments in threat assessment and management strategies. The simulated scenarios provide a controlled environment for testing response protocols, helping identify areas for improvement and reinforcing effective practices.

Guideline: Set clear training objectives based on real-world needs and conduct regular evaluations to assess progress. Adjust the curriculum according to field feedback to ensure that personnel are ready to meet evolving challenges.

Final Thoughts

In this chapter, we explored the essential skills and training methods that are vital for successful rescue operations. The focus was on developing readiness and efficiency among teams through structured and strategic training approaches. Key components included training in realistic scenarios to foster quick decision-making and seamless coordination under pressure. By practicing these techniques repeatedly, rescuers can build muscle memory and confidence, ensuring they respond swiftly and accurately during emergencies. Additionally, we highlighted the importance of honing specific competencies such as physical conditioning, risk assessment, technical tool mastery, and effective communication, all critical for adapting to dynamic rescue situations.

As we move forward, it's important to remember that continuous learning and practice are the backbone of effective rescue operations. Properly structured training programs, incorporating both theoretical knowledge and practical drills, help ensure that first responders remain adept and prepared for any emergency. This chapter serves as a guide to integrating strategic thinking into everyday training routines, aligning them with real-world needs. By maintaining these practices, rescue teams will be well-equipped to handle the demands of their roles, ultimately enhancing their ability to save lives and manage crises efficiently.

Chapter 10 Test Questions and Answers

Why is training crucial for the success of rescue operations?

Training is essential because it equips teams with the necessary skills, knowledge, and readiness to act swiftly, improving response times and effectiveness during emergencies.

How does training enhance response times in rescue operations?

Through regular training, teams become familiar with protocols, roles, and procedures, which allows them to execute tasks promptly without hesitation.

What role do realistic scenarios play in rescue training?

Realistic scenarios expose trainees to potential emergency conditions, helping them build confidence, develop adaptability, and practice decision-making under pressure.

What is the psychological benefit of realistic rescue simulations?

Realistic simulations help build mental toughness, prepare responders for the emotional challenges of real emergencies, and inoculate them against stress.

Why is effective communication important in rescue operations?

Clear communication ensures coordination among team members, reduces confusion, and helps prioritize tasks efficiently during high-pressure situations.

How can targeted decision-making training improve rescue operations?

It encourages critical thinking, quick evaluation of risks, and making informed decisions under stress, which optimizes the use of resources and improves effectiveness.

What is muscle memory, and how does it benefit rescue operations?

Muscle memory is the ability to perform tasks automatically through repeated practice, which enhances efficiency and reduces errors during high-stress rescue situations.

How does training under stress benefit rescuers?

Stress-based training prepares rescuers to perform effectively under pressure, reducing panic and improving calmness and decision-making in chaotic environments.

What are the four core competencies in rescue training?

The four core competencies are physical conditioning, rapid risk assessment and critical thinking, mastery of technical tools, and effective communication.

Why is physical conditioning important in rescue training?

Physical conditioning ensures rescuers have the strength, stamina, and flexibility to handle demanding situations, such as carrying heavy equipment or navigating tough terrain.

What role does critical thinking play in rescue operations?

Critical thinking allows rescuers to quickly assess risks, anticipate hazards, and make informed decisions, which can save lives during emergencies.

Why is mastery of technical tools crucial in rescue operations?

Mastery of tools like self-contained breathing apparatus (SCBA) and thermal imaging cameras ensures rescuers can safely navigate hazardous environments and locate victims efficiently.

How does effective communication improve teamwork during rescues?

Effective communication minimizes miscommunication, ensures clear roles, and enhances coordination, enabling teams to work together more efficiently during stressful operations.

What is the importance of incorporating regular evaluations into rescue training?

Evaluations help track progress, identify areas for improvement, and ensure that training is meeting the dynamic needs of rescue operations.

How does scenario-based training enhance decision-making skills in rescues?

Scenario-based training helps responders practice evaluating situations, making rapid decisions, and adapting to changing conditions, which improves their performance under pressure.

What is the Kirkpatrick Model, and how does it apply to rescue training?

The Kirkpatrick Model is a framework for evaluating training effectiveness through participant reactions, learning outcomes, behavior changes, and results, ensuring continuous improvement in rescue training.

Why are live drills and simulations important in rescue training?

Live drills and simulations provide hands-on experience and help rescuers practice skills in realistic conditions, preparing them for actual emergency situations.

How can simulation-based training with Virtual Reality (VR) benefit rescue operations?

VR simulations allow responders to practice decision-making and problem-solving in risk-free environments, enhancing their readiness without the constraints of physical drills.

Why is feedback integration important after live drills or simulations?

Feedback allows rescuers to reflect on their performance, identify areas for improvement, and continuously refine training techniques and protocols.

How can rescue teams improve their coordination during live drills?

By practicing clear communication, using effective teamwork strategies, and ensuring that all team members understand their roles, coordination can improve significantly during live drills.

Reference List

(2025). Fema.gov. https://emilms.fema.gov/is_0015b/groups/205.html

Anderson, G. S., Di Nota, P. M., Groll, D., & Carleton, R. N. (2020, October 20). *Peer Support and Crisis-Focused Psychological Interventions Designed to Mitigate Post-Traumatic Stress Injuries among Public Safety and Frontline Healthcare Personnel: A Systematic Review.* International Journal of Environmental Research and Public Health. https://doi.org/10.3390/ijerph17207645

Billings, J., Zhan, N., Nicholls, H., Burton, P., Zosmer, M., Albert, I., Grey, N., Sharif El-Leithy, Murphy, D., Tehrani, N., Wheatley, J., Bloomfield, M., & Greene, T. (2023, November 23). *Post-incident psychosocial interventions after a traumatic incident in the workplace: a systematic review of current research evidence and clinical guidance.* European Journal of Psychotraumatology; Taylor & Francis. https://doi.org/10.1080/20008066.2023.2281751

Emergency Response Services. (n.d.). Center for Disaster Philanthropy. https://disasterphilanthropy.org/resources/emergency-response-services/

Emergency Responder Experiences with the Incident Command System: A Qualitative Analysis - ProQuest. (n.d.). Www.proquest.com. https://search.proquest.com/openview/e659f6e2ffc7d0c002e0df7edfe0c166/1?pq-origsite=gscholar&cbl=18750&diss=y

FasterCapital. (n.d.). *The ongoing need for preparedness and collaboration in search and rescue operations.* Retrieved from https://fastercapital.com/topics/the-ongoing-need-for-preparedness-and-collaboration-in-search-and-rescue-operations.html

Lakshmi Narayanan, R. G., & Ibe, O. C. (2015). *Joint Network for Disaster Relief and Search and Rescue Network Operations.* Wireless Public Safety Networks 1. https://doi.org/10.1016/b978-1-78548-022-5.50006-6

Staff, F. E. (2002, June). *A STRUCTURED SYSTEM FOR FIRE INVESTIGATOR SAFETY - Fire Engineering: Firefighter Training and Fire Service News, Rescue*. Fire Engineering: Firefighter Training and Fire Service News, Rescue. https://www.fireengineering.com/firefighter-training/a-structured-system-for-fire-investigator-safety/

Werner, C. (2015, April). *Fire Technology: Using Drones In the Fire Service*. Firehouse.com; Firehouse. https://www.firehouse.com/technology/drones/article/12041104/drones-in-the-fire-service

What is a Dynamic Risk Assessment? A Guide with Examples. (2024, May 10). Bodytrak. https://bodytrak.co/en-us/news/dynamic-risk-assessment-guide/

(2025). Stjohn.org.nz. https://www.stjohn.org.nz/first-aid/first-aid-library/immediate-first-aid1/resuscitation/?srsltid=AfmBOoomVglcpKbclF5jKLAPeHvGi1KuRcpI0pS4qqsyK2dAvS2grjIV

CCOHS. (2023, March 2). *CCOHS: Hazard and Risk - Risk Assessment*. Www.ccohs.ca. https://www.ccohs.ca/oshanswers/hsprograms/hazard/risk_assessment.html

Crisis Communication Strategies for Emergency and Crisis Management Officials. (2021, June 14). Onlinewilder.vcu.edu. https://onlinewilder.vcu.edu/blog/crisis-communication/

Down, H. (2024, July 21). *6 Types of Workplace Hazards | EcoOnline*. EcoOnline. https://www.ecoonline.com/en-us/blog/6-types-of-workplace-hazard/

Drumm, W. (2022, August 24). *Scene Assessment in the Real World - First Response Rescue Blog*. First Response Training International. https://www.firstresponse-ed.com/blog/scene-assessment-in-the-real-world/

Nowak, T. (2024, October 2). *5 EMS safety considerations for structure collapse response*. EMS1. https://www.ems1.com/safety/articles/5-ems-safety-considerations-for-a-structure-collapse-response-JSukfiTNYLHLZeXg/

Penney, G., Launder, D., Cuthbertson, J., & Thompson, M. B. (2022, February 4). *Threat assessment, Sense making, and Critical decision-making in police, military, ambulance, and Fire Services.* Cognition, Technology & Work. https://doi.org/10.1007/s10111-022-00694-3

Structural Collapse Guide | Occupational Safety and Health Administration. (2025). Osha.gov. http://www.osha.gov/emergency-preparedness/guides/structural-collapse

Staff, J. (2022, September 20). *The Importance of a Coordinated Emergency Response Among Utilities and State and Local Governments.* Juvare. https://www.juvare.com/how-a-coordinated-emergency-response-can-saves-lives/

Dunn, S. (2024, January 19). *Rope Rescue Techniques.* Onropes.co.uk. https://www.onropes.co.uk/rope-rescue-techniques

Fire. (2019, February). *Technical Rescue Teams: Maintaining Staffing and Training - Fire Apparatus: Fire trucks, fire engines, emergency vehicles, and firefighting equipment.* Fire Apparatus: Fire Trucks, Fire Engines, Emergency Vehicles, and Firefighting Equipment. https://www.fireapparatusmagazine.com/training/technical-rescue-teams-maintaining-staffing-and-training/

Ford, A. (2023, April 17). *PPE Care Maintenance & Inspection Guide.* Zeroheightsafety.com; ZERO Height Safety (NZ/AU) Limited. https://zeroheightsafety.com/blog/care-maintenance-and-inspection-guide

Gislason, E. (2024, October 15). *Guide to Confined Spaces - Types of Confined Spaces | NASP.* National Association of Safety Professionals. https://www.naspweb.com/blog/guide-to-confined-spaces/

JIBC. (n.d.). *Chapter 3 – Incident Command System.* Pressbooks.bccampus.ca. https://pressbooks.bccampus.ca/esm1/chapter/chapter-11-incident-command-system/

McCullar, R. (2019). *Technical rescue: The importance of 15:1. Firehouse.* Retrieved from https://www.firehouse.com/rescue/rope/article/21095773/the-importance-of-151

No Ice Is Safe Ice – Ice Rescue Special Operations By Andrea Zaferes And Walt Hendrick - Lifeguard Systems, Inc. (2018, December 7). Lifeguard Systems, Inc. https://teamlgs.com/no-ice-safe-ice-ice-rescue-special-operations-andrea-zaferes-walt-hendrick/

TheDoctorFeelGood. (2025). *Rope Rescue Manual.* Scribd. https://doi.org/100154662/original/6959a2cf9d/1

Tabibian, R. (2025, January 9). *Comprehensive Machine Maintenance Safety Guide.* WorkTrek. https://worktrek.com/blog/machine-maintenance-report/

Working in Confined Spaces: Safety Guide [iAuditor]. (2018). SafetyCulture. https://safetyculture.com/topics/confined-space-safety/

Al-Qadami, E., Razi, M. A. M., Shah, S. M. H., Pu, J. H., Amran, M., Díaz Huenchuan, M. A., & Avudaiappan, S. (2024, July 14). *Floodwaters and vehicle hydrodynamics: A deep dive into risk mitigation unraveling vehicle stability in floods.* Results in Engineering; Elsevier. https://doi.org/10.1016/j.rineng.2024.102540

DIDIOKMAKINGS. (2025, February 11). *Rescue Techniques in Different Water Environments: Oceans, Rivers, and Floods.* DIDIOK MAKINGS. https://didiokmaking.com/blogs/news/rescue-techniques-in-different-water-environments-oceans-rivers-and-floods?srsltid=AfmBOoqCaUKUf6z-96hjCGw0dtAJzDJMf8JhxW2VEpnCaJGhyAQWZ11_

Fire Engineering Staff. (1995, May). *SWIFTWATER RESCUE TACTICS PART 1: LOW-RISK OPTIONS - Fire Engineering: Firefighter Training and Fire Service News, Rescue.* Fireengineering.com. https://www.fireengineering.com/technical-rescue/swiftwater-rescue-tactics-part-1-low-risk-options/

Hurst, A. (2024, January 24). *Swiftwater/ Floods and Training - Trident Rescue, USA.* Trident Rescue, USA. https://tridentrescue.com/swiftwater-floods-and-training/

McCabe, M. (2019, December 1). *Water-Rescue Techniques in the Cold Environment.* Firehouse. https://www.firehouse.com/rescue/water-ice/article/21112773/technical-rescue-water-rescue-techniques-in-the-cold-environment

Wang, L. (2022, November 1). *A review of the flood management: from flood control to flood resilience.* Heliyon. https://doi.org/10.1016/j.heliyon.2022.e11763

Building Collapse Rescue: Operational Considerations. (2019). Firehouse. https://www.firehouse.com/rescue/article/10544569/building-collapse-rescue-operational-considerations

Collapse Incident Marking Systems for Structural Triage and Rapid Search Reconnaissance Operations. (2002, November 4). Firehouse.com; Firehouse. https://www.firehouse.com/rescue/article/10573971/collapse-incident-marking-systems-for-structural-triage-and-rapid-search-reconnaissance-operations

Daley, M. (2018, May 1). *StackPath.* Www.firehouse.com. https://www.firehouse.com/rescue/technical-rescue/structural-collapse/article/20997907/building-collapse-basics-for-firefighters

Emergency Response Standard; Proposed Rule | Occupational Safety and Health Administration. (2024, February 5). Osha.gov. https://www.osha.gov/laws-regs/federalregister/2024-02-05

Federal Register :: Request Access. (n.d.). Unblock.federalregister.gov. https://www.federalregister.gov/documents/2024/02/05/2023-28203/emergency-response-standard

Hu, D., Li, S., Chen, J., & Kamat, V. R. (2019, October). *Detecting, locating, and characterizing voids in disaster rubble for search and rescue.* Advanced Engineering Informatics. https://doi.org/10.1016/j.aei.2019.100974

Lee, M. (2007, November 28). *Staying Safe at Structural Collapse Operations.* FireRescue1. https://www.firerescue1.com/rescue/articles/staying-safe-at-structural-collapse-operations-kzixo558Oge7eqeF/

Marcus. (2024, October 18). *How Search and Rescue Teams find survivors after earthquakes?* Fire Product Search. https://www.fireproductsearch.com/how-search-and-rescue-teams-find-survivors-after-earthquakes/

Structural Collapse Flashcards. (2025). Quizlet. https://quizlet.com/ca/21543246/structural-collapse-flash-cards/

Terry, M. (2017, August 21). *STRUCTURAL COLLAPSE AWARENESS.* Slideplayer.com. https://slideplayer.com/slide/6323117/

Augenstein, J., E Perdeck, Bowen, J., Stratton, J., Singer, M., Horton, T., Rao, A., Digges, K., A Malliaris, & J Steps. (2025). *Injuries in Near-Side Collisions.* Annual Proceedings / Association for the Advancement of Automotive Medicine. https://pmc.ncbi.nlm.nih.gov/articles/PMC3400216/

Bossert, R. (2024, July 22). *EMS: How New Automobile Technology Affects Assessment and Treatment.* Firehouse.com; Firehouse. https://www.firehouse.com/ems/article/55056745/how-new-automobile-technology-affects-ems-assessment-and-treatment

Extrication Cribbing & Stabilization. (2018). Industrial Protection Services, LLC. https://www.ipp-ips.com/Products/Extrication-Cribbing-Stabilization/

Ingram, M. (2023, June 13). *Trapped Under Trucks: The Inside Story of the Government's Failure to Prevent Underride Crashes.* ProPublica. https://www.propublica.org/article/underride-crashes-nhtsa-dot-iihs-safety-cars-trucks

Jon's Mid America. (2020, November 19). Jon's Mid America. https://jonsmidamerica.com/which-is-the-best-power-source-for-your-rescue-tools/

Loone, C. M. (2020, April). *To the Rescue: New Vehicle Extrication: What You Don't Know Can Kill You - Fire Apparatus: Fire trucks, fire engines, emergency vehicles, and firefighting equipment.* Fire Apparatus: Fire Trucks, Fire Engines, Emergency Vehicles, and Firefighting Equipment. https://www.fireapparatusmagazine.com/equipment/extrication-tools/new-vehicle-extrication-what-you-dont-know-can-kill-you/

Rescue tools: Pros and cons of 3 power sources. (n.d.). FireRescue1. https://www.firerescue1.com/fire-products/extrication-tools-cutters-and-spreaders/articles/rescue-tools-pros-and-cons-of-3-power-sources-2ME7grwCy5sMNiys/

TR VR STABILIZATION EXTRICATION Flashcards. (2025). Quizlet. https://quizlet.com/579947848/tr-vr-stabilization-extrication-flash-cards/

Colling, R. L., & York, T. W. (2010). *Emergency Preparedness—Planning and Management.* Hospital and Healthcare Security. https://doi.org/10.1016/B978-1-85617-613-2.00024-0

Ciesa, M., Grigolato, S., & Cavalli, R. (2015, June). *Retrospective Study on Search and Rescue Operations in Two Prealps Areas of Italy.* Wilderness & Environmental Medicine. https://doi.org/10.1016/j.wem.2014.10.008

Hawkins, S. C., & Simon, R. B. (2021, February). *Ten Myths about Medical Emergencies and Medical Kits.* Advances in Archaeological Practice. https://doi.org/10.1017/aap.2020.47

Marsteiner, J. (2024, January 26). *81 Basic Survival Tips That You Should Know.* Thesurvivaluniversity.com. https://thesurvivaluniversity.com/survival-tips/essential-survival-tips

Pittman, M. A., Slone, T., & Wilson, M. W. (2020, January 1). *Introduction to Wilderness Medicine-A Medical School Elective.* PubMed; National Institutes of Health. https://doi.org/10.21980/j8b93x

Survivor: The Ultimate Guide To Thriving In Extreme Situations. (2025, March 8). StarGaze. https://waf.inspectorsedge.com/trendi2/survivor-the-ultimate-guide-to-thriving-in-extreme-situations.html

UFO Themes. (2018, October 16). *Technical Rescue Interface: Search and Rescue in the Non-Snow/Glacier/Mountaineering Environment.* Anesthesia Key. https://aneskey.com/technical-rescue-interface-search-and-rescue-in-the-non-snowglaciermountaineering-environment/

Vincent-Lambert, C., Pretorius, A., & Tonder, B. V. (2023, December 1). *Use of Unmanned Aerial Vehicles in Wilderness Search and Rescue Operations: A Scoping Review.* Wilderness & Environmental Medicine. https://doi.org/10.1016/j.wem.2023.08.022

Wozniakowski-Zehenter, M. (2024, June 3). *Emergency Response Notification System and How to Use It Efficiently.* Identecsolutions.com; Identec Solutions. https://www.identecsolutions.com/news/emergency-response-notification-system-and-how-to-use-it-efficiently

vcsar1admin. (2017, September 28). *Rescuing Lost & Injured People in the Wilderness | VCSAR1.* Vcsar1.org. https://vcsar1.org/services/rescuing-lost-injured/

(2014). Fema.gov. https://emilms.fema.gov/is_0015b/groups/203.html

Clarkson, L., & Williams, M. (2023). *EMS, mass casualty triage.* PubMed; StatPearls Publishing. https://www.ncbi.nlm.nih.gov/books/NBK459369/

DeNolf, R. L., & Kahwaji, C. I. (2019). *EMS, Mass Casualty Management.* Nih.gov; StatPearls Publishing. https://www.ncbi.nlm.nih.gov/books/NBK482373/

FEMA. (2024, September 30). *National incident management system.* Www.fema.gov. https://www.fema.gov/emergency-managers/nims

Kuza, C. M., & McIsaac, J. H. (2018, December 1). *Emergency Preparedness and Mass Casualty Considerations for Anesthesiologists.* Advances in Anesthesia. https://doi.org/10.1016/j.aan.2018.07.002

Melmer, P., Carlin, M., Castater, C. A., Koganti, D., Hurst, S. D., Tracy, B. M., Grant, A. A., Williams, K., Smith, R. N., Dente, C. J., & Sciarretta, J. D. (2019, December). *Mass Casualty Shootings and Emergency Preparedness: A Multidisciplinary Approach for an Unpredictable Event.* Journal of Multidisciplinary Healthcare. https://doi.org/10.2147/jmdh.s219021

Pre-Hospital Mass Casualty Triage and Trauma Care | ASPR TRACIE. (2019). Hhs.gov. https://asprtracie.hhs.gov/technical-resources/33/pre-hospital-mass-casualty-triage-and-trauma-care/0

Ramesh, A., & Kumar, S. (2010). *Triage, monitoring, and treatment of mass casualty events involving chemical, biological, radiological, or nuclear agents.* Journal of Pharmacy and Bioallied Sciences. https://doi.org/10.4103/0975-7406.68506

A firefighter's guide to fireground search and rescue – Part 2. (n.d.). FireRescue1. https://www.firerescue1.com/fire-products/rescue-equipment/articles/a-firefighters-guide-to-fireground-search-and-rescue-part-2-TRLszcH6HvQpBhlJ/

Aumond, A. (2024, February 1). *Fire Safety in High-Rise Buildings: 3 Unique Challenges & Solutions | Performance Based Fire.* Www.pbfpe.com. https://www.pbfpe.com/post/high-rise-fire-safety

A firefighter's guide to fireground search and rescue – Part 1. (n.d.). FireRescue1. https://www.firerescue1.com/fire-products/rescue-equipment/articles/a-firefighters-guide-to-fireground-search-and-rescue-part-1-GuDRdPu3Yj8IZFxC/

Extreme Fire Behavior: Understanding the Hazard | CTIF - International Association of Fire Services for Safer Citizens through Skilled Firefighters. (n.d.). Ctif.org. https://ctif.org/news/extreme-fire-behavior-understanding-hazard

Firefighter Search and Rescue Manual & Tips 2021. (n.d.). Fire Training Toolbox. https://www.firetrainingtoolbox.com/firefighter-search-rescue/

Graham, J. (2023, September 10). *Navigating challenges: Fire safety management in high-rise buildings - PlanRadar - HR.* PlanRadar - HR; PlanRadar – HR. https://www.planradar.com/au/fire-safety-management-in-high-rise-buildings/

Staff, F. (2022, November 24). *Size-Up: Why It Is So Important.* FirefighterNation: Fire Rescue - Firefighting News and Community. https://www.firefighternation.com/training/size-up-why-it-is-so-important/

StackPath. (n.d.). Www.firehouse.com. https://www.firehouse.com/operations-training/article/10575457/rescue-and-squad-companies

Staff, F. E. (2018, November 19). *Size-Up and Risk Benefit for Small Commercial Structures - Fire Engineering: Firefighter Training and Fire Service News, Rescue.* Fire Engineering: Firefighter Training and Fire Service News, Rescue. https://www.fireengineering.com/firefighting/size-up-and-risk-benefit-for-small-commercial-structures-katherine-ridenhour/

Unlocking the Secrets of Fire Behavior: A Vital Evolution in Firefighting Tactics. (2024, January 23). https://www.fireengineering.com/firefighting/unlocking-the-secrets-of-fire-behavior-a-vital-evolution-in-firefighting-tactics/

Education - National Association For Search And Rescue. (n.d.). Nasar.org. https://nasar.org/page/Programs

Explorance. (2023, May 24). *5 steps to creating and implementing effective training programs.* Explorance. https://explorance.com/blog/5-steps-to-creating-effective-training-programs/

Express, L. S. (2024, January 27). *Emergency Preparedness Drills and Training | Life Safety Express.* https://lifesafetyexpress.com/life-safety/life-safety-articles/preparedness-drills-training/

List of Federal Financial Assistance Programs Funding Emergency Communications | CISA. (n.d.). Www.cisa.gov. https://www.cisa.gov/safecom/emergency-comms-grants-list

Piatt, L. (2023, May 8). *Overcoming Search and Rescue Challenges*. Rigging Lab Academy. https://riggingdlabacademy.com/overcoming-search-rescue-challenges/

RNR Rescue. (2023, September 14). *The Crucial Role of Training in Industrial Rescue - RNR Rescue*. RNR Rescue. https://rnrrescue.com/the-crucial-role-training-industrial-rescue/

SOLUTION. (2024, February 19). *The Kirkpatrick Model: 4 Levels of Training Evaluation*. Epilogue Systems. https://epiloguesystems.com/blog/the-kirkpatrick-model/

Syed Sarosh Mahdi, Hafsa Abrar Jafri, Raheel Allana, Gopi Batteneni, Khawaja, M., Syeda Sakina, Agha, D., Rehman, K., & Amenta, F. (2023, May 24). *Systematic review on the current state of disaster preparation Simulation Exercises (SimEx)*. BMC Emergency Medicine. https://doi.org/10.1186/s12873-023-00824-8

Training for any emergency – without the risk. (2025, March 5). AirMed&Rescue. https://www.airmedandrescue.com/latest/long-read/training-any-emergency-without-risk

Also by Eric J. Neal

From the Streets to a Chief
Igniting Excellence: Challenges of a First-Year Fire Chief
Echoes From the Ashes
The Roadmap to Fire Behavior
The Suppression Lesson: Water vs Fire
From the Shadows: SMU's Journey
Hazardous Materials Unveiled
Rescue Operations Strategies for Saving Lives
Firefighter Survival: Tactics for Staying Alive
Incident Command and Leadership Decision Making on the Fireground
Apparatus & Equipment Mastery Tools of the Trade
Smoke Between Us
Firefighter EMS & Medical Aid in the Fire Service
Clean Smoke & Dirty Silk
The Mercer Trap

About the Author

Eric J. Neal is a Fire Chief, Emergency Management Coordinator, educator, speaker, and author dedicated to developing leaders in the fire service. With decades of experience in public safety, he combines real-world leadership lessons with practical guidance to help current and future fire service professionals succeed.

In *Igniting Excellence: Challenges of a First-Year Fire Chief*, Eric shares firsthand experiences and valuable insights into the leadership, operational, and administrative challenges faced during a fire chief's first year.

www.ingramcontent.com/pod-product-compliance
Lightning Source LLC
LaVergne TN
LVHW022042200926
848481LV00003B/750

* 9 7 9 8 2 3 0 3 0 0 9 6 0 *